INFECTIOUS DISEASES

ENCYCLOPAEDIA OF ANIMAL DISEASES-V

INFECTIOUS DISEASES

By

Dr. Ashok Kumar
Dept. of Zoology
Bundelkhand University
Campus Department
Jhansi (India)

DPH

DISCOVERY PUBLISHING HOUSE PVT. LTD.
NEW DELHI-110 002

Reprinted – 2019
First Published - 2008

ISBN: 978-81-8356-206-5 (Set)

ISBN: 978-81-8356-286-7

Infectious Diseases

Published by:
DISCOVERY PUBLISHING HOUSE PVT. LTD.
4383/4B, Ansari Road, Darya Ganj
New Delhi-110 002 (India)
Phone: +91-11-23279245, 23253475; 43596065
E-mail: discoverybooksindia@gmail.com
discoverypublishinghouse@gmail.com
web: www.discoverypublishinggroup.com

Printed at:
Infinity Imaging Systems
Delhi

PREFACE

The **Infectious Diseases** has been carefully compiled and edited to meet the long felt needs of increasingly large number of those who have to deal with the different aspects of human diseases in colleges, universities and research institutes. It provides a stimulating and important new view of interaction between animals and pathogens causing diseases. The objective is to introduce to students the essential principles for understanding various aspects of diseases. Most of diseases constitute the largest part of human pathology and are the primary cause of death. Hence, special importance is given to the study of such diseases.

The book is intended to acquaint students of various fields involved directly or indirectly with the major principles of human diseases. The book may be helpful as well to practitioners and those engaged in medical research.

In the preparation of this book large number of books and research papers have been consulted. So no authenticity is claimed.

The author wishes to express his deepest appreciation to the many people who have contributed in one way or the other to the preparation of this title.

The author expresses his gratitude to Mr. Wasan and staff of M/s Discovery Publishing House for their whole hearted co-operation in the publication of this book.

The author tried hard to be accurate and upto date in statement and realises the impossibility of completely avoiding errors therefore, the author will greatly appreciate having his attention called to any questionable statement.

Author

CONTENTS

1

HANDLING AND PROCESSING OF SPECIMENS

A major concern of clinical microbiologists has been the rapid identification of significant microorganisms isolated from clinical specimens. The scientific literature contains many publications directed at the identification of genus, species, or both in the least possible time; yet the subject of this chapter has been almost totally ignored, despite the fact that it constitutes the foundation for all subsequent work. It is not possible to present in the space available an allencompassing guide to the procurement, proper management, and primary inoculation procedures of clinical specimens. Our interest is to present a working outline that is admittedly not all inclusive, but is an approach that knowledgeable clinical microbiologists should pursue in the handling of clinical specimens in their laboratories. There are many variations on the approach presented here.

It is necessary to establish certain ground rules for clinical microbiology. The purpose of the clinical microbiology laboratory is to rapidly and accurately provide the clinician with information concerning the presence or absence of a microbial agent that may be involved in an infectious disease process. When indicated, this information should be supplemented with an antibiotic profile. It is more important for a physician to know that a gramnegative rod susceptible to various antimicrobial agents is present than to wait several days or weeks for a properly identified microorganism. Clinical microbiologists must admit that, when identifying a microorganism, they base the identification on a number of salient characteristics. Clinical microbiologists are not taxonomists in the strict sense; they do not weigh each microbial character on an equal scale and, after collecting an appreciable number of characteristics, determine the identity of the organism according to Adansonian precepts. It is not surprising to find that clinical microbiologists use a variety of identification schemes that do not necessarily reflect the classification schemes of taxonomists or even follow *Bergey's Manual* of *Determinative Bacteriology. To* bridge any communication gap that may occur with the medical staff, the staff of the clinical microbiology laboratory should be aware of the synonyms of the organisms that are identified by the accepted genus and species designation. The approaches outlined in this chapter require the modifications deemed applicable by each clinical microbiologist for the needs of a particular laboratory. They, also must be modified to accommodate the existing requirements of prospective payment policies. A thorough knowledge of microorganisms expected in specimens from nonsterile body sites or common contaminants of sterile material, acquaintance with the recent history of the patient and the clinical impression of the physician, and the occurrence and antibiotic profiles of the more usual isolates involved in nosocomial complications must bear on decisions that affect the workup of specimens and the identification of the microorganisms they contain. Judicious decisions on the amount of analytical effort to be expended are now mandatory to provide high-quality microbiology for the patient in need of such care. The discriminate pruning of test procedures, elimination of unimportant procedures, rejection of unnecessary or improperly submitted specimens, and concerted educational effort in the ordering of tests will help to lessen the cost of health care.

COLLECTION OF SPECIMENS

Anatomical and Surgical Pathology

The microbiological examination of tissues obtained at the time of surgery or autopsy is of considerable importance because the specimen may represent the entire pathological process. Surgical specimens are obtained at considerable expense and'some risk to the patient. Further surgery to obtain more material may be contraindicated or refused; additional autopsy specimens often are not available because the remains have already been embalmed and buried. The microbiologist must be prepared to do whatever is necessary to establish the diagnosis, frequently with assistance from the pathologist. Many of the necessary procedures have been described in detail and form the basis of the discussion which follows.

Selecting the proper specimen and collecting an adequate sample for examination are essential. When the lesion is large or when there are several lesions, multiple specimens from different sites must be collected. Samples from an abscess should include pus and a portion of the wall of the abscess. The use of a cotton swab to collect a small amount of pus from an abscess cavity is tantamount to malpractice! It must be remembered that abscesses of the brain, lung, pleural space, peritoneal cavity, liver, pelvis, and wounds commonly contain anaerobes or a mixed flora and that an adequate sample of infected material should be collected and transported to the laboratory in an anaerobic container. When granules are observed in the pus, they should suggest infection with actinomycetes.

Gross surgical specimens submitted for histopathological examination are ideal for microbiological study since portions of the specimen may be carefully selected for analysis before placing the tissue in fixative. A close rapport between microbiologist and surgical pathologist is invaluable. Histological examination reveals whether or not the lesion is malignant and, if it is inflammatory, whether it is granulomatous or suppurative. If the lesion is malignant, it may not be necessary to perform cultures. If it is inflammatory, the type of reaction may indicate what types of culture are necessary; however, it is well to bear in mind that the histopathology of infections due to mycobacteria, fungi, and brucellae is quite variable, but the diseases are not mutually exclusive. Histopathological diagnosis of an infection depends on the visualization of a sufficient number of organisms in characteristic form. Tissue stains for acid-fast bacilli provide positive results in only 30 to 40% of those specimens with positive cultures, whereas special stains for fungi often show organisms not recovered in cultures, or they may show structures resembling the organisms but lacking definitive characteristics. Cultures, therefore, are absolutely necessary to establish or confirm the diagnosis of an infectious process.

A gross surgical specimen may be bisected aseptically by the surgeon. One half is submitted to surgical pathology and the other half is placed in a sterile, wide-mouthed, screw-capped jar and sent to the microbiology laboratory. Lung biopsy tissue impressions are made by imprinting the freshly cut surface several times on each of two or three glass slides for staining with Giemsa, Gomori methenamine-silver, or toluidine blue for *Pneumocystis carinii. The* tissue is finely minced with sterile scissors and ground aseptically with a pestle in a sterile mortar, with 60-mesh aluminum oxide (alundum) as an abrasive. Alternatively, tissue may be homogenized in a sealed plastic bag by using a Stomacher Lab Blender. The lung homogenate may be centrifuged and the sediment used to prepare smears for *P. carinii.* This procedure yields improved detection of *P. carinii* cysts compared with impression smears. Slides for Gram and special stains may be prepared from the homogenate to determine the presence of microorganisms and guide the culture approach. Antigen detection methods may also be applied if appropriate and if reagents are available. A 20% suspension of the ground material is made in broth and placed in a sterile dropper bottle to be used for inoculating culture media and for storage. Preparing and examining special histological stains may take several days, and further microbiological studies also may be indicated; therefore, it is advisable to store any residual tissue emulsion at 5°C for 1 or 2 weeks before discarding it. It is recommended that as much material as possible be examined microbiologically by the inoculation of multiple plates or tubes containing the appropriate media. Inoculation of appropriate cell cultures for virus isolation should be done if indicated. If it is anticipated that virus isolation may be attempted later, some of the body secretions or tissue (whole or ground) should be stored at - 70°C. If there is any intention to consider electron microscopy study of the tissue for viruses, a portion of the involved tissue should be preserved in glutaraldehyde.

It is generally thought that cultures of embalmed tissue are useless; however, Weed and Bagenstoss were able to isolate tubercle bacilli, *Histoplasma capsulatum, Nocardia asteroides,* and various species of bacteria from tissues that had been embalmed for 24 to 48 h. Embalming will diminish considerably the probability of recovering an organism; however, in certain cases, cultures of tissues embalmed for as long as 48 h may be worthwhile, provided nothing else is available and the area of the embalmed tissue selected for cultures is centrally located.

The value of postmortem microbiology remains controversial. In some institutions the heart blood is routinely cultured; in others, cultures are limited to cases and specific organs in which an infectious disease is suspected. In recent years, attempts have been made to determine whether clinical or autopsy evidence of infection is an appropriate mechanism for selecting cases to culture, whether sampling should be limited to tissues or organs suspected of being infected, and whether the microbiologist should attempt to correlate postmortem culture results with those obtained antemortem. In general, the results of these studies have demonstrated that humans possess an indigenous tissue flora, that there does not appear to be postmortem transmigration of organisms, that the frequency of positive cultures is poorly related to clinical or autopsy evidence of infectious disease, that there is a substantial lack of correlation between ante- and postmortem culture results, that cultures performed on a single postmortem tissue only are rarely of value, and that in some cases postmortem cultures of multiple tissues may be of value in identifying the etiological agent of the infection, especially if these represent cases with wellrecognized clinical entities caused by a single organism or cases with an overwhelming infection. The value of postmortem cultures is, therefore, limited except in selected cases.

When postmortem cultures are indicated, the prosector should obtain at least a 6-cm^3 portion of tissue, with one serosal or capsular surface intact. The tissue should then be placed in a sterile, sealed, plastic container and sent to the laboratory for immediate processing or storage at 5°C. Subsequent procedures are based on those described by Dolan. In the laboratory, the capsular or serosal surface is thoroughly seared with a soldering iron and incised with a sterile instrument. A 1-cm^3 portion of tissue is removed aseptically from the core of the tissue block; it is used to prepare impression smears and is then ground, as described above, to provide a 10 to 20% emulsion to be used as inoculum for cultures. When a small specimen is received, it may be immersed in boiling water for 3 to 5 s to decontaminate its surface. Any residual portion of surgical or anatomical specimens should be retained in a refrigerator by the microbiologist until it is reasonably well established that there is no longer any need to retain the tissue.

The routine described by Isenberg and Berkman for postmortem examination has also been found useful by a number of pathologists and entails qualitatively surveying lungs, liver, spleen, and kidney, as well as obtaining heart blood. Smears and cultures are made of the material aseptically obtained at the time of autopsy. Comparison between the findings from the direct smears and especially the inoculated blood plates usually indicates correlation. One of the major points which must be stressed is the need to obtain microbiological cultures from, areas grossly appearing infected or involved in an infectious process. The past history of the patient, the course of therapy, and the agonal events must also be considered so that media to culture organisms that do not grow readily on routinely used media can be obtained. Such examinations may serve also to delineate the most significant nosocomial microbes in the environment of a specific hospital.

Blood Cultures

The prompt and accurate isolation and identification of the etiological agents of septicemia remain among the most important functions performed by the clinical microbiology laboratory. The indications for obtaining blood cultures are a sudden relative increase in the pulse rate and temperature of the patient, a change in sensorium, and the onset of chills, prostration, and hypotension. Other indications include a prolonged, mild, and intermittent fever in association with a heart murmur. Bacteremia is continuous in endocarditis, endarteritis, uncontrolled infections, typhoid fever, and brucellosis, and it is usually intermittent in other infections. Timing in the collection of cultures in endocarditis, for example, may not be critical; however, in other cases, timing of the collection is important because the bacteremia is usually intermittent and may precede the onset of fever or chills by as much as 1 h.

In patients with suspected bacterial endocarditis, three blood cultures are sufficient to isolate the etiological agent in nearly all instances. These should be collected separately and, the condition of the patient permitting, at no less than hourly intervals within a 24-h period. In intermittent bacteremias, three separate blood cultures within 24 to 48 h are usually sufficient to isolate the etiological agent. In such instances, the time interval between cultures is frequently determined by clinical circumstances and the urgency to initiate antimicrobial therapy. In patients who have received antimicrobial agents before blood collection, a total of four to six separate blood cultures may be necessary to isolate the etiological agent, or consideration should be given to using special techniques to adsorb or inactivate the antibiotic(s) present in the blood.

It is essential that blood for culture be collected aseptically, first by cleansing the skin with 80 to 95% alcohol and then by applying 2% iodine in concentric fashion to the venipuncture site. Because of the lower incidence of skin hypersensitivity to iodophors, they may be used in lieu of iodine. Instant antisepsis never occurs, and the iodine or iodophor should remain intact on the skin for at least 1 min. The intended venipuncture site should not then be touched unless the fingers used for palpation are similarly disinfected. After the vehipuncture, any residual iodine shoul.d'be removed with an alcohol sponge or pad.

It is recommended that 10 to 20 ml of blood be collected for each culture. Culture of a lesser volume may result in lower recovery rates because of the relatively low numbers of organisms present in most bacteremias. In infants and children, collection of 1 to 5 ml appears to be satisfactory. Optimally,. the blood is inoculated directly into culture media at the bedside of the patient, with either a syringe and needle or a transfer set. Alternatively, the blood may be transported to the laboratory in a sterile, evacuated tube containing sodium polyanetholesulfonate (SPS) and then inoculated into culture media. Several new approaches have become available recently and are summarized by Reller. These may prove to be useful substitutes for the traditional methods and may decrease the time-to-positive interval as well as the technical attention requirements.

It is essential that blood be inoculated into culture media on a 10% (vol/vol) basis to counteract the normal bactericidal activities of chemical and cellular mediators of immunity. Any residual bactericidal effects remaining after this dilution of blood in the culture media have been shown to be abolished by the presence of 0.05 and 0.025% SPS, which is a polyanionic anticoagulant that is also anticomplementary and antiphagocytic.

Any general-purpose, commercially available nutrient broth medium may be used for the culture of blood. Soybean casein digests, such as tryptic soy broth, Trypticase soy broth, Columbia broth, and brain heart infusion broth, have been found to be satisfactory. Strict reliance on fluid thioglycolate, Thiol, or supplemented, prereduced anaerobically sterilized media is not recommended because of lower isolation rates of aerobic or facultatively anaerobic bacteria in these media. Commercially available liquid media are generally bottled under vacuum with CO_2 and contain 0.025% SPS. As such, they are satisfactory for cultivation of anaerobes from blood. Bottles containing 50 or 100 ml of medium should be employed. The value of liquid media made hyperosmotic, generally by the addition of sucrose, remains uncertain. In general, such media have not been found to increase the detection rates or to decrease the time interval for detection of bacteremia.

Because of the mutually exclusive atmospheric requirements of several groups of microorganisms encountered in septicemia, it is desirable to use two bottles, one which is not vented and one which is vented. Venting is performed by aseptically inserting a sterile, cotton-plugged needle through the rubber stopper of the bottle and then withdrawing the needle after the vacuum within the bottle has been released. If the venting unit is left in place, the bottle should be incubated in an atmosphere containing 10% CO_2.

Cultures are incubated at 35°C and are examined macroscopically or radiometrically later on the day of their collection and daily thereafter for at least 7 days for evidence of growth. Clinically significant bacteria are recovered within this period in at least 95% of instances; however, longer periods of incubation may be necessary with specimens from seriously ill patients who are receiving and appear not to be responding to antimicrobial agents or from patients with endocarditis or endarteritis presumably due to fastidious microorganisms. Recovery of such bacteria beyond the first 7 days of incubation may provide the opportunity for administration of more specific and effective antimicrobial agents.

Gram- or acridine-orange-stained smears and aerobic and anaerobic subcultures of obvious or suspected positive cultures should be prepared immediately, and the results of the microscopic examination of the smear, if positive, should be reported by phone and in writing to the physician of record as quickly as possible. Subcultures for direct antibiotic susceptibility testing may be performed; however, these results should be considered tentative and require confirmation by retesting the isolate with a standardized method.

Routine (blind) subcultures of grossly negative cultures should be performed between 6 and 14 h after blood collection and after 48 h of incubation. Subcultures are inoculated onto quadrants of chocolate blood agar plates which are incubated in 3 to 6% CO_2 for 48 h. Subcultures are necessary to ensure the recovery of *Pseudomonas aeruginosa, Haemophilus* spp., meningococci, gonococci, and yeasts, particularly when blood culture bottles are not vented. Routine anaerobic. subcultures are unnecessary. In some instances a routine Gram- or acridine-orange-stained smear made within 24 h after blood collection has been helpful.

The recovery of diphtheroids (aerobic and anaerobic), *Bacillus spp.*, and *Staphylococcus epidermidis* usually signifies contamination, unless they are present in multiple cultures. Nonhemolytic streptococci, excluding group D streptococci, and alpha-hemolytic streptococci in single cultures are of uncertain significance. *Bacteroidaceae, Enteroba-*

cteriaceae, P. aeruginosa, Haemophilus spp., Staphylococcus aureus, pneumococci, and yeasts are nearly always clinically significant.

Polymicrobial bacteremia has been reported in as many as 18.4% of microbiologically proven septic episodes and has been associated with higher mortality than has unimicrobial infection.

The routine addition of penicillinase to blood culture media does not appear to be justified, except in selected patients who are receiving high doses of a penicillin or cephalosporin at the time of blood collection. In such instances, it is essential to prepare concurrent cultures of the penicrlinase solution, contamination of which has been associated with outbreaks of pseudosepticemia. In general, antimicrobial agents are inhibited or neutralized by dilution of the blood in broth, addition of SPS to the medium, and lysis-concentration techniques. Although the advantages of antibiotic-adsorbent resins are controversial and may be system dependent, it is generally agreed that such devices should always be used in conjunction with conventional blood culture systems.

The isolation of brucellae and many other organisms is facilitated by inoculation of the blood into Castaneda double medium or into soybean casein digest broth, which must be subcultured at least twice weekly for 4 weeks. In either case, prolonged capneic incubation is required. Castaneda double medium without SPS is useful in recovering pathogenic neisseriae from blood.

Leptospiremia is usually present only during the first week of illness. One to three drops of freshly drawn blood are inoculated into each of several tubes containing 5 ml of a suitable semisolid medium, such as Fletcher or Ellinghausen medium, and the cultures are incubated in the dark at 30°C for 28 days. A portion of each culture is examined weekly by darkfield or fluorescence microscopy.

Although it has been recommended that detection of fungemia should be accomplished by inoculation of blood on a 10% (vol/ vol) basis into a soybean casein digest broth with subcultures to fungal isolation media at 48-h intervals or by inoculation of the blood into a Castaneda double medium, it now appears that fungal isolation rates are substantially improved by lysis-concentration techniques.

Bone

Specimens from orthopedic procedures or postoperative complications must be transported to the laboratory quickly and in media which prevent drying. Microbiologists should insist that each culture be accompanied by a smear procured from the same site as the culture. It is important to insist that information on the patient be provided to permit analysis for various microorganisms not included in the routine screening of surgical specimens. Thus, specimens from animal bites which penetrated to the periosteum require inoculation of media which support *Haemophilus haemoglobinophilus* and *Pasteurella multocida.*

Special care and consultation with the surgeon of record must be exercised with traumatic and nosocomially complicated osteomyelitides. Microbiologists must do their utmost to help the surgeon differentiate between skin organisms and those from the deeper recesses of tissues and bone. Multiple specimens, obtained with great care and accompanied by smears, can help provide this information. Blood cultures are important adjunct procedures in monitoring such patients.

Ear

The bacteriological findings from persons with acute otitis media have been sufficiently consistent in many reported studies that there would seem to be little justification for performing tympanocentesis, except in therapeutic failures or in neonates, in whom the bacterial etiology differs from that of older children. Since the correlation between nasopharyngeal and middle ear cultures is poor, cultures of the naso-pharynx may provide misleading results. Most studies show that *Streptococcus pneumoniae,*

Haemophilus influenzae, and *Streptococcus pyogenes* are the most common etiological agents of acute otitis media; however, *Branhamella catarrhalis, P. aeruginosa, Escherichia coli, Klebsiella pneumoniae,* and *S. aureus* may also be isolated from middle ear effusions of infants, and in therapeutic failures S *aureus* and *P. aeruginosa* are predominant.

Eye

In patients with conjunctivitis and, especially, with keratitis, there are special problems in specimen collection and processing. Swabs are generally inadequate for establishing the presence of microorganisms because of the small sample size. An additional problem is the antimicrobial activity of topical anesthetics. Microbes responsible for keratocoznjunctivitis include bacteria, actinomycetes, fungi, and viruses. It is, therefore, recommended that swabs for cultures be taken before topical anesthetics are applied and that corneal scrapings be taken after they are applied. Ideally, a laboratory technologist should assist the ophthalmologist in collecting the specimen and preparing the cultures.

Examination of Gram- and Giemsa-stained smears of the corneal scrapings may provide preliminary clues to the nature of the disease. If necessary, stains for acid-fast bacteria or a potassium hydroxide wet mount can be made. Because of the limited amount of material available, it is recommended that the scrapings be spotinoculated directly onto each one of the following media: chocolate blood agar, brain heart infusion agar with 5% sheep blood, inhibitory mold agar, and, if indicated, Lowenstein-Jensen agar. Swab cultures may be inoculated onto one half of each agar plate and the scrapings onto the other half.

Feces

Proper collection and preservation of feces is a frequently neglected but important requirement for the isolation of microorganisms contributing to intestinal disease. Unless the specimen can be taken immediately to the laboratory and properly handled upon delivery, a number of important microorganisms will not survive the changes in pH which occur with a drop in temperature. This is especially true of most shigellae and an appreciable number of salmonellae. When delays are unavoidable, it should be a standard rule that all stool specimens be submitted in a stool preservative, such as 0.033 M phosphate buffer mixed with equal volumes of glycerol. An indicator may be added which will assure an approximate pH of 7. The indicator added to the specimen provides visual proof that the pH drop has been inordinate. Stool specimens are introduced into a screw-capped glass container with stool preservative immediately after they have been passed. A 0.5- to 2-g quantity is sufficient. If sterile swabs are used in obtaining the specimen, they should be passed beyond the anal sphincter, carefully rotated, and withdrawn. The swabs may be added to a screw-capped tube containing preservative and transported to the laboratory for culture. Difficulties are occasionally encountered when patients have diarrhea and are unable to use a bedpan. Under those circumstances, paper towelling may be deposited into the toilet bowl, and the diarrheal stool may be obtained above the water level. Toilet paper, which usually has been impregnated with bismuth salts in the manufacturing process, should not be used, since a number of organisms found in feces are inhibited or killed by these salts.

If the patient is hospitalized, personnel obtaining the specimen should be instructed explicitly to choose portions of the stool that display either mucus, blood, or both. These areas usually harbor a large number of the organisms that are involved in the disease process.

It cannot be stated too often that the vast majority of bacteria found in feces are anaerobic gram-negative rods. They are usually ignored in the attempt to diagnose infectious intestinal disorders. This is not to imply that such organisms may not be involved in disease processes of the intestinal tract. However, to date, such a causal relationship has not been established, with the exception of food-borne *Clostridium perfringens* intoxication and pseudomembranous colitis after antimicrobial therapy resulting in the implantation of *Clostridium difficile.* It is important to remember that the biochemical activities of the anaerobic microbiota contribute to the detriment of *Enterobacteriaceae.* Therefore, the facultatively anaerobic enteric bacteria require the type of buffering provided by a stool preservative.

There are times when surveys for carriers as well as institutional outbreaks indicate a need to use rectal swabs to obtain specimens. Under these circumstances, the stool preservative should not be used; the swab should immediately be placed, instead, into a medium such as Gram-Negative broth (BBL Microbiology Systems), especially if the responsible organism has already been identified and contacts or carriers are being sought. Enrichments such as tetrathionate broth or selenite F may be used, particularly if salmonellae are involved.

As with most other specimens, a single stool specimen with little or no accompanying information is inadequate. The laboratory can be much more helpful to the clinician if a brief history is provided. If necessary, the physician of record should be called for this information so that the clinical microbiologist can be particularly attentive to culturing the stool on media most likely to yield the responsible bacteria, fungi, or parasites. Without such information, procedures cannot be modified to accommodate as complete a microbiological analysis as is required. Of equal concern are the increased cost and the time lost in searching for impossible organisms.

A single negative stool culture cannot be regarded as sufficient for laboratory confirmation of noninvolvement of infectious bacteria. Although the number three is not inviolate, repeated cultures are indicated if the clinical picture suggests the involvement of bacteria and the first two cultures are unrewarding despite the symptoms. Similarly, after the diagnosis has been made, microbiological surveillance of the convalescent individual and of contacts who may have become carriers should be conducted at regular intervals until at least three negative specimens have been obtained consecutively.

When the history or early examination suggests the possibility that anaerobes, *Bacillus sp.*, or staphylococci; *Vibrio cholerae, Vibrio parahaemolyticus*, and related bacteria; *Campylobacter sp.;* toxigenic or invasive *E. coli;* and certain viral agents may be involved, special media should be used for direct inoculation of the stool. Specimens collected for the demonstration of parasites and ova are discussed in the section on parasitology. Gram-stained smears for the demonstration of bacteria have not been used frequently in recent times. In the past, examination of such smears was popular for determining the ratio of gram-negative to grampositive bacteria. It is not possible to determine the type of microorganisms present simply by observing the morphology and staining characteristics of bacteria in a stool specimen. The usefulness of such a method is restricted to cases such as pseudomembranous colitis, staphylococcal enterocolitis, or monilial disease, where the overwhelming number of gram-positive cocci or yeastlike cells would easily be demonstrated in a properly made Gram stain of a portion of the stool specimen or rectal swab; extraordinary numbers of pus cells can be detected in this fashion also.

Body Fluids

Cerebrospinal fluid

The examination of cerebrospinal fluid (CSF) from patients suspected of having meningitis represents one of the major emergency procedures faced by personnel in the clinical microbiology laboratory. The reasons for this urgency are that bacterial meningitis is a rapidly fatal disease if untreated or inadequately treated and that appropriate antimicrobial therapy often requires prompt identification of the etiological agent. Lumbar puncture and examination of the CSF should be undertaken whenever the physician suspects meningitis or wants to rule it out. It must be remembered that the typical signs of meningeal irritation in the adult, such as fever, headache, vomiting, nuchal rigidity, hyperreflexia, etc., are usually absent in infants and neonates, in whom the clinical manifestations of meningitis are often vague and nonspecific. An unexplained febrile illness in an irritable infant who is doing poorly should lead one to suspect meningitis. Meningitides due to mycobacteria, fungi, leptospires, or protozoa are generally insidious in onset. The diagnosis of viral meningoencephalitis is frequently established by exclusion and by serological means. Although some viruses may be isolated from CSF, viral isolates are more likely to be obtained from other clinical specimens.

Lumbar puncture must be performed under conditions of strict asepsis, since contamination of the specimen can occur readily and confuse the identification of the etiological agent. The skin should be disinfected with povidone-iodine. Specimens should be collected in sterile containers which can be sealed with a screw cap to preclude leakage and loss or contamination of the. contents. Cotton-plugged or rubber-stoppered tubes should not be used, and snap-top containers should be checked to ensure that a tight seal does occur and that some of the contents are not aerosolized on opening. The sterility and absence of microorganisms in CSF specimen containers should be periodically confirmed by culture and also presumptively by Gram stain. Methods for the detection of microbial antigens in CSF are indicated now that several types of reagents are available commercially. These approaches are recommended especially if no organisms can be detected microscopically or to confirm the identity of the common etiological agents seen on smears or both. Prompt transport of the specimen to the laboratory is mandatory, since fastidious organisms such as H. *influenzae* and *Neisseria meningitides* may not survive storage or variations in temperature. For these reasons, some advocate -that smears be prepared and cultures be inoculated at the bedside of the patient when the CSF is obtained. Though such a practice may be ideal, it is seldom practical. The clinical microbiologist, therefore, should examine the lumbar puncture tray routinely used in the hospital to ensure that the CSF containers are of satisfactory quality (not all commercially available trays have airtight specimen containers), (ii) attempt to establish a standardized skin preparation, and (iii) develop the systems whereby the specimen can be transported promptly to the laboratory. An adequate sample. (as much as possible) of CSF should be available for microbidogical examination, particularly when a diagnosis of tuberculous or fungal meningitis is considered, since the numbers of microorganisms present are often small. If only one specimen container is filled, it should be submitted to the microbiology laboratory first so that it can be opened aseptically and samples for chemical and cytological studies can be removed at the time cultures are inoculated. The microbiology laboratory should process the fluid immediately by preparing Gram-stained smears and inoculating the appropriate media.

Since usually only small numbers of microorganisms are present in infected CSF, some procedure for the concentration of any organisms present should generally be performed. The simplest of these procedures is centrifugation of the specimen at 1,000 x g fore 15 min; then the supernatant fluid is removed for chemical or serological studies, and the sediment is used for both smear and culture purposes. This centrifugation is adequate for those fluids in which there is an increase in inflammatory cells; however, its limitation is that in very early cases of bacterial meningitis and in fungal meningitis the cell count may be normal despite a positive culture. Furthermore, a force of 10,000 x g for 10 min is required to sediment *H. influenzae,* which means that with most standard clinical laboratory centrifuges, which develop a maximum of 1,000 x g, 60 min may be necessary. For these reasons, prolonged (30-min) examination of the Gram-stained smear under oil immersion may be necessary. The same sampling problem occurs when attempts are made to culture the specimen from a nonexistent sediment. In instances in which there is no cellular response or a poor one, it is recommended that the specimen be concentrated instead by filtration through a 0.45-μm membrane filter which is directly placed on an appropriate culture medium. Sterile, disposable, 13- and 25-mm filter units are satisfactory and convenient for this purpose.

A variety of immunological procedures have been developed for the rapid diagnosis of bacterial meningitis; however, standardized reagents and some special equipment are requird for some of these procedures. Fluorescence microscopy, capsular swelling, counterimmunoelectrophoresis, coagglutination, and latex particle agglutination are some of the principal techniques used. The usefulness of these tests is in those instances in which the patient has received antibiotic therapy and in life-threatening situations when a fast and reasonably presumptive diagnosis is required. These tests should be well controlled to be reliable. In most laboratories, however, the Gram-stained smear remains the most reliable and convenient rapid diagnostic tool.

The inflammatory and noninflammatory responses of the host may be helpful in the differential diagnosis of meningitis. Whereas the leukocytic response in the CSF in acute bacterial meningitis is usually polymorphonuclear, that in tuberculous, fungal, leptospiral, or protozoan meningitis is usually lymphocytic and less intense. Although polymorphonuclear leukocytes may be predominant in the CSF early in the course of aseptic meningitis, there is usually a clear shift to mononuclear cells within 8 h. The CSF glucose is usually depressed in cases of acute bacterial or tuberculous meningitis and normal in the other meningitides; CSF protein is usually elevated. Cytological and chemical changes in the CSF may occur in patients with brain abscess; however, smears and cultures of the CSF are generally negative in these cases unless the abscess ruptures into the subarachnoid space or the ventricles.

One final point in partially treated cases of bacterial meningitis deserves emphasis, and that is that cellular and chemical findings and the recovery rates of bacteria in cultures may be altered. Furthermore, a tendency has been noted for gram-positive organisms to appear to be gram negative in such cases, so that findings in a Gram-stained smear must be interpreted cautiously.

When fungal meningitis is suspected, a drop of the CSF sediment should be mixed with a drop of India ink or nigrosin solution on a clean glass slide, covered with a cover slip, and examined with a decreased intensity of light. Nigrosin is preferable to India ink because it is free from discernible particulate matter. The presence of encapsulated, budding, yeastlike cells in the wet preparation is virtually diagnostic of cryptococcal meningitis; however, special care must be exercised in differentiating nonencapsulated yeasts from erythrocytes or leukocytes, air bubbles, or even talc. In some cases of crypttococcal meningitis, there is little or no cellular response or depression of the sugar value in the CSF.

Cryptococcal meningitis is frequently associated with disseminated infection, and it is not uncommon to recover *Cryptococcus neoformans* from sources other than CSF (urine, blood, sputum, bone marrow, etc.). In patients from whom *C. neoformans* has been recovered initially from such other sources, serious consideration must be given to lumbar puncture and examination of the CSF, since symptoms referable to the central nervous system may be absent. In patients with suspected cryptococcal meningitis, performance of the latex-cryptococcal antigen test with CSF is often helpful because the India ink or nigrosin preparation is negative in roughly 50% of instances.

The development of meningoencephalitis of obscure etiology in patients with a recent history of swimming in stagnant fresh water before the onset of symptoms should lead to the prompt examination of the CSF for motile amoebae.

Fluids other than CSF

As with the lumbar puncture, the percutaneous aspiration of pleural, pericardial, peritoneal, and synovial fluids must be performed aseptically to avoid contamination of the specimen and to prevent the accidental introduction of microorganisms into these anatomical spaces. The specimen should immediately be injected into a sterile tube or bottle. Since infection of these spaces may be due to anaerobes, it is recommended that fluid or pus be collected with a sterile syringe and needle, that any air bubbles present in the syringe be expelled, and that the material then be injected into an anaerobic transport tube or vial containing CO_2. Again, as much material as it is feasible and practical to collect should be submitted to the laboratory. A small amount of sterile heparin may be added to the fluid to prevent coagulation, since clots may trap microorganisms. Coagulated material should be emulsified and cultured along with a portion of its surrounding fluid. The special procedures and media advocated for the isolation of *Legionella pneumophila* should be applied to pleural fluids.

Gram-stained smears of the centrifuged sediment of clear or slightly cloudy fluids should be examined carefully; however, frankly purulent material should be smeared directly and examined after it has been stained for the presence of bacteria.

Respiratory Tract

Cultures of the respiratory tract must be interpreted cautiously because of the microflora normally present in the nose, oral cavity, and pharynx and because of the frequency of nosocomial acquisition of potentially pathogenic microorganisms by seriously ill patients. Since potential pathogens such as S. *aureus, H. influenzae, Streptococcus pneumoniae, P. aeruginosa, Enterobacteriaceae,* and yeasts may be present in the oropharynx, their isolation from cultures of the respiratory secretions does not represent a priori evidence of their etiological role in respiratory infections.

Nasopharyngeal cultures should be performed to detect carrier states of *N. meningitidis, Corynebacterium diphtheriae,* and *S. pyogenes*. In addition, such cultures aid in the diagnosis of whooping cough and croup. Nasopharyngeal specimens should be obtained with a Dacron, cotton, or calcium alginate swab on a flexible wire which is gently passed through the nose into the nasopharynx, rotated, removed, and placed into a suitable transport medium for isolation. Nasopharyngeal aspirates are superior to swabs for isolating *Bordetella pertussis* from children suspected of having whooping cough.

Although nasopharyngeal cultures are sometimes recommended in comatose patients unable to expectorate a sputum sample, the frequency of aspiration and subsequent anaerobic pleuropulmonary infection in these patients and the misleading results provided by oropharyngeal contamination make percutaneous transtracheal aspiration mandatory.

H. influenzae type B may cause a form of croup known as acute epiglottitis. The distinctive appearance of the epiglottis and the rapidly progressive and fulminating course, which may lead to death within 24 h, demand prompt initiation of therapy. Nasopharyngeal and blood specimens should be obtained and cultured for *H. influenzae* once an airway has been assured.

It should be kept in mind that orolaryngeal involvement is not uncommon in both acute and chronic disseminated forms of histoplasmosis and blastomycosis, in tuberculosis, and in leishmaniasis. Cultures for these agents and biopsy with histological demonstration of the organisms are necessary to establish the correct diagnosis. It is imperative that the clinician communicate his or her clinical suspicions to the microbiologist immediately. Otherwise, valuable time and specimens may be lost.

Throat cultures are obtained most frequently for the diagnosis of streptococcal pharyngitis and less commonly for the diagnosis of pertussis, diphtheria, and pharyngitis due to gonococci or viruses. There is no need to identify other microorganisms in routinely submitted throat cultures, since there is little or no evidence to document their role in producing pharyngitis. Exudative pharyngitis, enlarged cervical nodes, headache, nausea, vomiting, and abdominal pain are commonly associated with streptococcal pharyngitis, whereas cough, rhinorrhea, and hoarseness are commonly associated with viral pharyngitis. There are, however, many viral infections which are confused clinically with streptococcal pharyngitis. Acute tonsillopharyngitis with vesicles or shallow ulcers on the anterior fauces, palate, and buccal mucosa is usually due to herpes simplex virus or to coxsackievirus A. A history of an incomplete immunization series or no immunization during childhood should alert the clinician to the possibility of infection with *B. pertussis* or *C. diphtheriae,* particularly when incomplete immunization is coupled with the characteristic signs and symptoms of these diseases. Gonococcal pharyngeal infection should be suspected in patients with gonococcal infections at other sites, especially among those practicing fellatio.

Requisitions for culture or specimens from the upper respiratory tract should specify the suspected etiological agent so that laboratory personnel can take appropriate steps to ensure its isolation. The culture should be obtained under direct visualization with a Dacron, cotton, or calcium alginate swab by vigorously swabbing both tonsillar areas, the posterior pharynx, and any areas of inflammation, ulceration, exudation, or capsule formation. The tongue should be depressed with a tongue blade or spoon to minimize contamination of the swab with oral secretions which may dilute, overgrow, or inhibit the growth of pharyngeal flora.

Bacterial culturing of sputum is fraught with error, and clearcut results are seldom obtained. Specimens are frequently collected haphazardly by paramedical personnel who are not aware of the necessity for a fresh, clean specimen resulting from a deep cough and who fail to transport the specimen to the laboratory promptly. Expectorated sputum is frequently contaminated with oropharyngeal flora, and it is difficult to determine which of the many different potential pathogens isolated is responsible for pulmonary infection. Rarely, a potential pathogen is isolated in pure culture and may be presumed to represent the etiological agent. To quote Barrett-Connor, who found no pneumococci in 45% of the sputum and nasopharyngeal cultures examined from patients with pneumococcal bacteremia, "The routine sputum culture for the diagnosis of acute bacterial pneumonia may be a sacred cow. Not only can the results lead to serious mismanagement of the patient, but also sputum cultures represent one of the largest workloads and expenses for the hospital bacteriology laboratory." Correlation between results of cultures of transtracheal aspirates and results of qualitative and quantitative cultures of the sputum has been poor, except in patients producing sputum containing fewer than 25 squamous epithelial cells per x 100 field microscopically. There is, therefore, no completely satisfactory method for isolating bacterial pathogens from expectorated sputum at the present time. Under no circumstances is anaerobic bacterial culturing of sputum performed. A Gram-stained preparation should precede all sputum culture efforts. Fewer than 25 epithelial cells per lowpower field or at least equal numbers of pus cells and epithelial cells, preferably with a predominance of neutrophiles, are an absolute requirement of sputum specimens for cultural analyses.

Lower respiratory secretions collected by nasotracheal aspiration with a catheter may also become contaminated with upper respiratory flora. Another method, which bypasses oropharyngeal contamination completely, is percutaneous transtracheal aspiration, a relatively harmless procedure that can be performed rapidly with local anesthesia. This procedure is recommended for seriously ill. or comatose patients who have pneumonia and cannot raise sputum or who are suspected of having anaerobic pleuropulmonary infection. Serious complications are rare when an experienced and qualified physician performs this procedure; nonetheless, patients must be followed carefully. Since the saline itself may be bactericidal, any specimen so collected should be cultured promptly; if this is not feasible, a balanced salt solution, such as lactated Ringer solution, should be used instead. Specimens should be transported promptly to the laboratory or injected into an anaerobic transport tube or vial. It is recommended that a Gram-stained smear of the material collected by transtracheal aspiration be examined and that aerobic bacterial cultures be performed with all possible speed. In some instances, culturing for anaerobic bacteria, mycobacteria, and fungi may also be indicated. Direct staining of such aspirates with specific fluorescent antibodies may hasten the diagnosis of diseases such as legionellosis; a search for *P. carinii* may also be indicated, provided the appropriate clinical impression has been communicated to the laboratory staff.

Culture of early-morning freshly expectorated sputum or of expectorated sputum induced by a heated aqueous aerosol of 10% glycerol and 15% sodium chloride, followed in about 1 h by a gastric washing, is useful for recovery of mycobacteria and fungi. Gastric washings obtained after induced coughing may have more mycobacteria. A series of such collections may be desirable. Ideally, the specimens should be cultured on the day they are collected. Pooled specimens are sometimes collected for culture; however, to prevent overgrowth of contaminants, it is necessary to refrigerate each specimen until the pool is processed. Twentyfour-hour sputum collections are unnecessary and, in fact, *undesirable*.

In some instances, the etiology of pulmonary infections may be diagnosed by percutaneous transthoracic needle biopsy of the lung, thoracentesis, needle aspiration of an abscess or empyema cavity, or open lung biopsy. When open biopsy is performed, multiple specimens should be obtained from different sites if the lesion is large or if multiple lesions are present. In addition, a portion of an abscess wall should be removed, as well as a sample of the pus within. Histological examination serves two purposes: to determine whether the lesion is inflammatory and, if so, to provide clues as to the nature of the infectious agent. These clues may, in turn, lead to further histological studies with special stains and to additional special cultures. Open lung biopsy may be necessary to establish the diagnosis of infection due to *P. carinii* or *L. pneumophila* in immunocompromised hosts, and it may be helpful in either establishing or confirming infections due to mycobacteria, fungi, and viruses. In experienced hands, the diagnosis of pneumocystosis may be made rapidly by appropriately staining material obtained by bronchial or transbronchial biopsy.

Specimens obtained by bronchoscopy have generally not been suitable for bacterial culture because during the procedure the lower respiratory tract is inevitably contaminated with flora indigenous to the upper respiratory tract. It has been suggested that by inserting a distally occluded catheter with telescoping cannulas it is possible to minimize bacterial contamination of upper respiratory origin. At this time, it appears that the results of quantitative cultures of the protected catheter brush are more clinically useful than those obtained from qualitative cultures; however, it would appear prudent to culture bronchial material for bacteria on a selective basis only and to restrict anaerobic cultures to special situations.

Skin

Caution must be exercised with specimens alleged to originate from skin. In the absence of lesions that can be cultured readily, these specimens are of little help, at best. Frequently they are misleading, especially if the request does not specify examination for dermatophytes. In that instance, proper scrapings should be obtained. Routine cultures of cellulitis or rashes are usually of little help. Specimens from various and sundry lesions must be obtained after the area has been properly disinfected. Cultures and smears as well as media for virus isolation should be inoculated, especially if the skin lesions are closed and aspiration with a sterile syringe and needle can be performed by a qualified clinician. When dealing with productive lesions, surface material should be discarded, and exudates from the interior should be used for microbiological analysis. Submission of smears of such material should be mandatory. Microbiological monitoring of burns requires identification of the usually polymicrobic microbiota and should be accompanied by frequent blood cultures, especially if the injury is extensive.

Urine

Urinary tract infections encompass diseases that involve areas ranging from the kidney to the urethra, with the urethra and bladder most commonly affected. Infections are classified as uncomplicated (medical) when there are no anatomical or neurological abnormalities present and as complicated (surgical) when residual inflammatory changes, obstructive uropathy, calculi, or neurological lesions are present. Infections usually arise via the urethra by the ascending route and, less commonly, hematogenously. In most hospitals, urinary tract infections represent the most common form of nosocomial infection.

Females of childbearing age seen for the first time with symptoms of uncomplicated urinary tract infection are nearly always infected with *E. coli*. These strains are usually susceptible to antimicrobial agents, including sulfonamides, which are active against gram-negative bacilli. Because a high degree of correlation exists between urinary sterilization and in vitro susceptibilities, some argue that it is unnecessary to obtain cultures from these patients. However, the prevailing viewpoint is that it is essential to obtain urine cultures during therapy. According to Stamey et al., "Better medicine is practiced, at less expense to the patient, if a culture is repeated 48 to 72 hours after starting therapy than if a useless antimicrobial agent is continued for ten days in the face of ineffective therapy". Urine cultures obtained during therapy should be sterile, since the symptomatic response of the patient is a poor indicator of successful therapy. Pyuria is not a totally reliable index of the presence of disease because it may be absent during the process or may occur in disorders other than bacterial infection, such as extreme dehydration, trauma secondary to instrumentation or calculi, chemical inflammation, renal tuberculosis, acute glomerulonephritis, and nonbacterial gastroenteritis and respiratory infections.

The reliability of a culture of a single, clean-voided specimen is about 80% in the female and virtually 100% in the adult male if he is circumcised or has carefully retracted the foreskin and cleansed the glans. In females, the reliability of a clean-voided specimen increases to 90% and nearly 100% if two and three specimens, respectively, are obtained and contain the same organism. Obviously, in symptomatic patients one specimen only is usually collected before therapy is initiated; however, in asymptomatic patients two or three specimens should be collected to document the presence of bacteriuria.

Guidelines for the diagnosis of urinary tract infections and minimal diagnostic criteria have been presented in detail by Kunin. In most females, the clean-voided midstream method of specimen collection is satisfactory, provided that the patient is capable of cleansing herself and collecting her own specimen or that the nursing personnel, in the case of the bedridden patient, clearly understand printed or stated instructions. All too frequently, instructions are given casually. The result is a grossly contaminated specimen. In addition, the task of obtaining a specimen from the bedridden patient is frequently assigned to the most inexperienced hospital aide or orderly. Hospitals should be encouraged to establish urine collection teams or select individuals who are trained in the proper methods of urine collection and transport.

Urine collection for culture by urethral catheterization is seldom indicated, except in those cases in which catheterization must otherwise be done for diagnostic or therapeutic reasons. There is a small risk of infection after urethral catheterization which varies according to the type of patient catheterized. Because of normal urethral colonization with bacteria, it may be difficult to determine whether organisms isolated from a catheterized urine sample are of urinary or urethral origin.

Contamination of the urine by urethral or introital microflora may be obviated by suprapubic aspiration. This is a rapid, safe, and simple technique when performed by a qualified physician. Suprapubic aspiration is indicated in patients with clinical evidence of urinary tract infection but in whom bacterial counts in clean-voided specimens are low and, therefore, indeterminate, in neonates, in young infants, in patients in whom catheterization may be contraindicated, and in those with suspected anaerobic bacteriuria. The patient should have a full bladder at the time the procedure is performed. After the skin has been properly disinfected, a 19- or 20-gauge needle attached to a syringe is passed through the skin in the midline at a point approximately one-third the distance from the symphysis pubis to the umbilicus. Urine is aspirated into the syringe.

In patients with chronic indwelling urethral catheters attached to closed drainage, urine is collected for culture by disinfecting with a suitable agent the wall of the catheter at its juncture with the drainage tube and puncturing it with a 21-gauge needle attached to a syringe into which the urine is aspirated. It is possible to puncture such a catheter repeatedly without leakage occurring from the puncture sites. The connection between the catheter and the drainage tube should not be broken for specimen collection, nor should material for culture be taken from the drainage bag.

During the course of cystoscopy, ureteral catheterization, or retrograde pyelography, urine may be collected for culture. In the case of an obstructed ureter, bladder urine may be sterile, whereas the urine proximal to the obstruction may be infected; hence, the urologist will often collect several specimens and request that each be handled separately.

Transport and storage of urine specimens are important adjuncts to the reliability of culture results. A sterile, screw-capped container or tube should be used. Urine is an excellent culture medium, and a small or insignificant number of bacteria can multiply rapidly to a significant number unless certain precautions are taken. If the urine cannot be cultured within 1 h after its collection, it can be refrigerated with satisfactory results for at least 24 h and, according to one report, for as long as 5 to 10 days. The wisdom of prolonged refrigeration has recently been questioned, and refrigeration should be used only if absolutely necessary.

Significant bacteriuria occurs when there are 100,000 colonies or more per ml in a clean-voided, midstream specimen obtained from asymptomatic patients. However, for acute dysuria and frequency in young, sexually active women, a colony count of 100 CFU/ml may be a useful criterion. In such cases, it is imperative to notify the laboratory so that procedures appropriate to' the detection of as few as 100 CFU/ml are used. It must be emphasized that this criterion should not be extrapolated to populations that have either a substantially greater likelihood of contamination of voided urine specimens or a lower prevalence of infection. Lower counts in a catheterized specimen require reculturing for confirmation of the results. Any count in a suprapubic aspirate is significant, although recovery of skin contaminants, such as diphtheroids and staphylococci, may not be significant and may require repetition of the procedure to confirm or rule out their presence in the urine. It must be recognized, however, that bacteriuria may be low on a diurnal basis when the patient is well hydrated, if the urine pH is below 5.0 or its specific gravity is less than 1.003, in the presence of an antimicrobial agent, with complete ureteral obstruction, and in chronic pyelonephritis. The role of cell wall-defective variants in urinary tract infections has not been established clearly. Anaerobic bacteriuria, although not uncommon in clean-voided, midstream specimens, has been only rarely confirmed by suprapubic aspiration, which represents the only reliable means by which to establish this diagnosis.

Chemical and other newer screening tests notwithstanding, significant bacteriuria may be established with roughly 90% accuracy by microscopic examination of a Gram-stained smear of wellmixed, unspun urine or of a wet preparation of unstained urinary sediment. The presence of at least two bacteria per × 1,000 microscopic field of the Gram-stained smear signifies bacteriuria of 100,000 or more colonies per ml.

Candiduria is abnormal; its quantitation is of little importance and may even be misleading if standards currently applicable to bacteriuria are applied. What is important is its relationship to the clinical picture, and it may serve not only as a warning of generalized candidiasis but also as an indication of the presence of an underlying disease. *Candida spp.* are frequently grown on conventional bacteriological media; however, their recovery is improved, particularly when they are present in small numbers mixed with bacteria, on fungal agar media containing antibiotics, e.g., inhibitory mold agar. A microscopic examination of a wet preparation of urinary sediment may serve as easily and inexpensively to demonstrate the presence of yeasts and pseudohyphal elements.

Pyuria without bacteriuria indicates the possibility of renal tuberculosis. There may be gross or microscopic hematuria, and secondary bladder involvement may produce symptoms of cystitis. Three consecutive clean-voided, early-morning specimens should be collected for cultures. Twenty-four-hour urine collections are undesirable because of the frequent overgrowth of bacteria other than mycobacteria.

Finally, a urinary calculus should be split so that cultures can be made of its interior, as well as of its surface, since it is not uncommon to isolate bacteria from within the stone which are not present on its surface.

Venereal Disease

Until recently, concern with gonorrhea has dominated the field of venereal disease. At this time, the only diagnostic test is the microbiological demonstration of viable gonococci in the various exudates and lesions which are present in persons with this disease. In the male, the microscopic observation of diplococci resembling gonococci in a Gram-stained smear of a urethral discharge remains unproblematic. There is excellent correlation between the presumptive diagnosis of uncomplicated gonorrhea in the male based on the microscopic examination of a Gramstained smear of urethral exudate and the confirmation of the diagnosis by the culture and identification of the organisms by. standard techniques. On occasion, individuals attending a clinic or consulting their physicians may exhibit a purulent discharge with little or no evidence of microorganisms. In these instances, the discharge or the urethral canal should be sampled for culture inoculation. Direct inoculation of Thayer-Martin (TM) medium or modified TM medium is advisable in situations where an inoculated plate can be immediately transported to the laboratory. If the history and symptoms suggest exposure but only a small urethral discharge can be obtained, urethral scrapings should be obtained by the clinician with a small swab or a smooth, small wire loop; the secretions should be cultured appropriately, and a smear should be examined. If it is known or suspected that the patient is homosexual, oropharyngeal and rectal cultures are indicated. TM agar or modified TM agar plates should be inoculated directly. Recent reports of cryptic carriage of gonococci in the distal urethra of clinically asymptomatic men exposed to infected women make the culturing of this site with bacteriological loops inserted into the urethra for a short distance a significant test.

Obtaining an adequate specimen in female patients may present greater problems. The presence of a cervical discharge is helpful .in the diagnosis. However, the Gram-stained smear obtained from such a discharge cannot be relied upon to establish the absence or presence of the gonococcus, and cultures should be made without exception. A cervical culture and, when possible, an anal culture have been recommended. If pelvic examination suggests that the vaginal glands or the urethra is involved, swabs from these areas should also be cultured for the gonococcus on modified TM medium. Should transport to the laboratory be delayed, modified TM medium may be employed.

Skin lesions, especially in the female, have been observed with increasing frequency: Very often they appear at the same time as pudendal involvement. Cultures of scrapings from the base of such lesions have been successful, but they are positive in a comparatively low percentage of the cases. Careful uprooting of the skin lesion with a sterile needle should be employed for obtaining specimens from these sites. Staining of smears of material from such sites with *Neisseria gonorrhoeae* fluorescent-antibody conjugate is often helpful in establishing a rapid, presumptive diagnosis. The report should state how this presumptive diagnosis was determined.

When a diagnosis of gonococcal arthritis is suspected, properly obtained specimens should be cultured on a blood agar plate and a noninhibitory plate of chocolate agar. Gas chromatography of synovial fluid appears to be a rapid and useful technique in the laboratory diagnosis of gonococcal and other arthritides. Caution must be exercised before needling such joints, with particular emphasis on adequate skin preparation. It is advisable that smears be made from the aspirated material and that it be cultured also for organisms other than the gonococcus. The use of hypertonic media for the detection of cell wall-deficient variants is advisable. Finally, when gonococcemia is suspected, the blood culture bottles should be subcultured minimally at 1-, 2-, and 4-day intervals on noninhibitory chocolate agar.

Nonselective enriched chocolate agar, TM medium, or modified

TM medium is advocated for cultures of the exudate in ophthalmia neonatorum. Endocervical cultures of the mother should also be obtained immediately.

The immediate inoculation and proper incubation of the culture medium should be stressed by the clinical microbiology laboratory. It is comparatively easy to indoctrinate clinic and hospital personnel in the proper fashion of carrying out this procedure. Illustrative materials, available from the Centers for Disease Control, are most helpful in presenting the proper procedures. The microscopic observation of bacteria on smears prepared carelessly is often obscured by the thickness of the material allowed to dry on a slide. Ward and clinic personnel should be reminded continuously that thinner smears are more helpful, instructed how to prepare slides in a proper fashion, and shown the difference between thick and thin smears for the purposes of recognizing gonococci.

It is also necessary to instructt and inform physicians, nurses, and attendants that the gonococcus can remain viable for some time in certain body fluids and that precautions to protect themselves and laboratory personnel handling the material are required. Containers contaminated on the outside, swabs not properly transported in enclosures, and smears dripping with discharge should not be handled in the laboratory without all individuals using every precaution to prevent personal contamination.

Dark-field demonstration of *Treponema pallidum* from primary and secondary lesions of syphilis or the demonstration of *T. pallidum* with specific direct immunofluorescent conjugate are the direct microbiological examinations that may be requested of a clinical laboratory. These procedures are useless if performed on material obtained and transported to the laboratory. Patients with such lesions must be examined within the vicinity of a dark-field microscope. Proper preparation of these lesions, the preparation of the slide, and the recognition of the treponemes are not within the scope of this chapter. However, it should be evident that, because of the infrequency with which this examination is performed in the routine clinical microbiology laboratory, only very experienced personnel should be assigned the task of identifying *T. pallidum* in the darkfield preparations made from lesions. It may even be advisable to send patients to a laboratory where the test is performed routinely. The misdiagnosis of syphilis can be a traumatic and damaging experience to the patient and must be avoided at all costs.

Chancroid is rarely reported. Recently, Hammond et al. have shown that the isolation of *Haemophilus ducreyi* from lesions can be readily accomplished by applying a swab sample to supplemented chocolate agar containing 3 μg of vancomycin per ml. Proper attention to incubation is necessary.

Wounds, Boils, Abscesses, etc.

Material from a previously undrained abscess, if properly collected and transported to the laboratory, should contain the etiological agent(s) of disease in most instances. An opened wound, ulcer, or sinus tract, however, frequently becomes contaminated with skin, mucosal, or airborne microorganisms. In general, the use of a swab to collect material from these sites is of limited value, since the amount of material supplied for examination is not only very small but also likely to represent an inadequate sample. Moreover, nothing is more worthless than a dry swab from a dry lesion. In chronic, localized lesions the number of organisms present may be small. A sterile needle and syringe should be used to collect a generous quantity of liquid material. Since anaerobic bacteria are commonly recovered from certain wounds, such material should be sent to the laboratory immediately or, preferably, injected into an anaerobic tube or vial for transport. Since sinus tracts often originate in bone or lymph nodes, the orifice of the tract should be cleaned thoroughly with a suitable antiseptic; then curettings of the lining of the tract should be taken as close to the base of the tract as possible. Ulcerative lesions of the skin and mucosa should be biopsied or, if they are small, excised for histopathological and microbiological studies. Irrigation, intravenous, or intraarterial catheters should be carefully removed after their entry site in the skin is disinfected; the tips should be cut off aseptically and submitted in a sterile container to the laboratory for culture.

Repeated cultures of open draining wounds or of large areas of devitalized tissue will frequently yield a profusion of microorganisms. Their significance is uncertain, their numbers and variety tax the laboratory, and their presence is unlikely to be affected by rational systemic antimicrobial therapy. Repetitive antimicrobial susceptibility testing of isolates from such cultures not only wastes the laboratory's time and the patient's money, but may also provide misleading results, inasmuch as attempts at therapy may simply favor superinfection by resistant microorganisms.

Although acute infections are generally recognizable clinically and their etiology is usually established by conventional bacteriological procedures, the recognition and establishment of the etiology of chronic localized infections are not so readily accomplished. Whereas *staphylococci, streptococci, Enterobacteriaceae,* and *Pseudomonadaceae* predominate in the former type of infection, actinomycetes, brucellae, mycobacteria, and fungi must be considered in the differential diagnosis of the latter. As we have already emphasized, biopsy and histopathological examination are valuable adjuncts in the diagnosis of the latter type of infection. The importance of microscopic study of the tissue is emphasized further by the occasional resemblance between neoplasms and infectious processes.

Burn wounds represent a somewhat unique problem because they are initially contaminated with staphylococci, streptococci, and clostridia and subsequently with such opportunists as *Pseudomonas spp., Klebsiella spp., Providencia spp., Candida albicans, zygomycetes,* and *viruses.* The burn eschar is a nonviable, devascularized structure which provides a milieu favorable for microbial growth. Quantitative cultures of biopsies of black, degenerated, or unhealthy areas of the burn or of drainage from burn wounds may be helpful in establishing the etiology of local or invasive infections. Gram-stained smears of any drainage may be helpful in providing an early clue to the nature of the infectious process. Such patients must be monitored with blood cultures at frequent intervals.

Quantitative smears and cultures of tissue from traumatically acquired wounds, taken after surgical debridement, are useful in predicting the risk of wound sepsis and, therefore, in helping the surgeon decide whether or not to close the wound primarily.

BRINGING SPECIMENS TO THE LABORATORY

Inefficiency in transporting specimens to the laboratory after they have been obtained from the patient is a major problem and becomes a convenient excuse of laboratory workers for not finding organisms whose presence is suspected by the clinician. When a specimen is lost, the loss becomes a convenient excuse for the clinician who has left the specimen with allegedly responsible personnel who, in turn, have instructed others to transport it to the laboratory. Although this activity is not directly under the jurisdiction of a clinical microbiology laboratory, every effort must be made to control its operation. The best possible way to obtain acceptable results is to insist that transportation time be as short as possible. Some laboratories refuse to accept specimens if they have been in transport too long. Certain specimens should be transported in a medium or vehicle which preserves the organisms in the specimen and helps to maintain the ratio of one organism to the other in the specimen. This is especially important for those specimens in which normal microbiota may be admixed with bacteria or other microorganisms foreign to the location. It becomes even more necessary to use a transport medium if significant microorganisms are present in very low numbers. House medical staff and nursing supervisors should remind appropriate personnel as often as required that prompt and proper transport of specimens from the clinics or patient floors to the laboratory is important. Many laboratories have established a policy of refusing to process specimens that have been handled improperly.

Microbiological specimens may be transported to the laboratory by various means. A variety of media may be used directly for this purpose. A nonproliferating, buffer type of transport medium such as that of Stuart, Toshach, and Patfula may be employed; however, the charcoal transport medium is preferred by some. Ordinary nutrient broth or anaerobic broth may be used when swabs or aspirates are involved. In other instances, microbiologists insist that, for example, throat cultures for streptococci be inoculated just as the swab is obtained onto blood agar plates and that the inoculations be made near the periphery over an area ca. 2 cm in diameter. The swab used for obtaining the specimen is then broken into a fluid transport medium and submitted in conjunction with the blood agar plate. Modified TM medium is recommended for gonococcal specimens when they are to be transported to a laboratory. For the isolation of *B. pertussis,* direct inoculation of Bordet-Gengou agar is advocated. Many laboratories do not accept stool specimens for culture unless they are transported in various buffered preservatives. Preparation of these transport media is described in various manuals. The major points advocated in this manual may be summarized as follows. (i) Aspirated specimens, such as fluids, etc., should be collected with a sterile needle and syringe instead of with swabs which permit rapid drying and subsequent loss of viability of the bacteria. Such specimens should be transported promptly to the laboratory or injected into an anaerobic transport tube or vial. (ii) Microorganisms survive better if 0.5% agar is added to media such as thioglycolate, chopped meat, etc. (iii) Smears for staining should be prepared at once, especially when the material is from pathological lesions which have not been aspirated with a syringe and needle and from which a large quantity of material cannot be obtained. It is important for adequate and proper smears' to be prepared at the bedside by clinicians who fully understand the significance of this procedure and are aware that thick smears are of very little help. Thin smears can be prepared from material that seems thick and tenacious by firmly pressing two slides together over the areas containing the applied specimen. This method is recommended for urethral discharges, wound exudates, aspirated material from abscesses, etc. By simply sliding the slides apart, thin areas of smear are made which are useful as a guide for the clinical microbiologist in the selection of media for primary culturing.

Special treatment is required for specimens when the etiological agent is thought to be anaerobic. The material is aspirated with a needle and syringe. It may be practical to remove the needle, cap the syringe with its original seal, and bring the specimen directly to the laboratory. No more than 10 min ought to be allowed for the transport of such specimens; otherwise, such specimens must be injected immediately into an anaerobic transport tube or vial. The laboratory must be alerted that a specimen for anaerobic cultures is on the way. If prereduced media are available, they must be inoculated as soon as the specimen is obtained. It is the responsibility of the microbiology laboratory to ensure that personnel using this approach fully understand the method of inoculation. Swabs for anaerobic culture are usually less satisfactory than aspirates, even when transported in an anaerobic tube.

Transport of urine specimens to the laboratory presents some problems. The urine must be obtained properly and transported and processed as soon as possible. No more than I h should elapse between the time the specimen is obtained and the time of incubation. If this time schedule cannot be followed, the urine specimen must be refrigerated immediately. If transport to the laboratory takes more than a few minutes, the urine specimen should be placed in an insulated bag containing scotch ice. Upon delivery to the laboratory, the urine specimen should be cultured immediately or refrigerated again. In any event, it is not advisable to refrigerate urine for longer than 18 to 25 h before culture under routine circumstances.

All laboratory personnel must be aware that submitted material which is not in a supportive environment must be handled immediately, especially spinal fluid, urine, pus, or any material not transported in a preservative. All smears should be fixed and stained immediately. In addition, laboratory personnel doing the primary culturing of specimens should acquaint themselves with the clinical diagnosis and any other pertinent laboratory findings, especially with select specimens such as blood, spinal fluid, wounds, burns, etc. Thus, hematological and chemical data, immunochemical findings on levels of immunoglobulins, complement, etc., may frequently be helpful in interpreting the findings of the laboratory and planning for the isolation of certain specific microorganisms. Laboratory personnel must also remember to handle all clinical specimens and cultures with appropriate attention to laboratory safety for themselves and coworkers.

RECORDING THE DATA

The person collecting the specimen is responsible for ensuring that the name, registration, and location of the patient are correctly (and legibly) written or imprinted on the culture request form and that this information corresponds with that written or imprinted on a label affixed to the specimen container. The type or source of the sample and the physician's choice of tests to be performed must be specified on the request form. Errors in specimen identification may have disastrous consequences and must be avoided. Correct identification of the specimen, however, continues in the laboratory, where different types of cultures, subcultures, and other appropriate tests and procedures may proliferate and, ultimately, all results are compiled into one or more reports. Multiple specimens from the same patient may be collected for diagnostic purposes or for following the progress of therapy; for these specimens, sample source, timing, and type of test performed must be clearly recorded.

In most laboratories, a system of accession numbers expedites record keeping. There are undoubtedly many variations of the system at all levels of sophistication, depending on whether human or machine systems are employed. Since off- or on-line computer systems are still relatively uncommon in clinical microbiology laboratories, most accessioning systems are human and, therefore, subject to an incidence of random errors from 1 per 100 operations to 1 per 1,000 operations.

One approach to accessioning is the use of sequentially numbered self-adherent labels produced with sufficient replicates so that one can be affixed to each copy of the culture request form, the specimen container, work card(s), each of the primary plates or tubes of media inoculated, and the average number of plates or tubes of media used in subcultures. Further categorization according to the type of specimen and the year it was received can be achieved by colorcoding the labels and using a numbered prefix to indicate the year of accession, followed by one or two letters to designate the type of specimen received. A list of accessions by laboratory and'specimen type is helpful in documenting whether or not the specimen was received and can be used for rapid location and tracking of work in progress. Such a listing may also be used to tabulate the monthly and annual workloads received by the laboratory. These figures enable the laboratory director to anticipate requirements for technical assistance and space.

In addition to the specimen records discussed above, the dates and times the specimen was collected and received should be recorded on the request forms. If the specimen was received in an unsatisfactory condition, this fact should be noted in the accession list, on the request forms, and in an incidence record. The clinician of record must be notified by phone and in writing. Compliance with these standards not only is required by laboratory inspection and accreditation agencies but. also provides data that are useful in the delivery of proper patient care.

Most laboratories use some type of work card on which a record of the work in progress is noted and updated as each step is completed. If the work card format is systematized, an individual who is unfamiliar with a particular culture can answer inquiries about it before the final report is ready. Inspection of the work card to determine the appropriateness of the tests performed and the accuracy of their interpretation is an important responsibility of the laboratory director when the final report is prepared.

Final reports must be retained in the laboratory at least until the results are recorded in the history of the patient and in keeping with legal and accreditation requirements. Retention beyond this point is not practical unless it fulfills a specific purpose, be it epidemiological, investigational, or medico-legal, since record storage is expensive; retention of records is unnecessary if the data are also a part of other permanent records. Microbiological results provide the data base for the epidemiology program' of a hospital and should be organized in a manner which facilitates analysis of infection rates by medical or surgical services and hospital location. This organization will enhance detection of hospital endemics or epidemics. Antimicrobial susceptibility data should be tabulated by organism and by antimicrobial agent on a yearly basis to detect trends in susceptibility and provide clinicians with the information needed for selecting initial therapy for serious infections. Several different computer systems address these problems. Unfortunately, experience to date is still too limited to warrant discussion or recommendation.

Many variations of specimen recording are extant. The essential features of the approach described must be part of all recordkeeping efforts. It should be stressed that the record must reflect, in addition, the various steps which comprised the analytical process, to ensure accuracy and detection of aberrant reactions while enabling other laboratorians to follow the decision-making process at a later date. All quality controls exercised in the interest of the particular specimens must be recorded. Although it may not be an absolute requirement that each specimen record reflect this concern, the various controls exercised must be logged to provide the reassurance of proper materials, reagents, media, etc.

PRIMARY CULTURING

A variety of considerations must be entertained by the technical personnel when any specimen is delivered to a laboratory: the source of the specimen, the transport medium or status of the specimen, thee accompanying smear or the nature of the specimen from which a smear is to be made, and the established routine for the laboratory in general. All of these matters must be brought into play for the technical personnel to determine the steps to be taken in preparing specimens for primary culture.

The section on Collection of Specimens (above) lists a variety of anatomical sources from which laboratory specimens may be obtained. It seems reasonable that in any laboratory all of the specimens should be cultured on a certain series of microbiological media. Which media will be used in any given laboratory is left to the experience of the director or supervisor of the clinical microbiology laboratory. The consideration of possible anaerobic microorganisms in a particular specimen must be accompanied by the use of media capable of supporting growth of anaerobic bacteria. Another important factor is the antimicrobial therapy history of the patient from whom the specimen was obtained. The physician submitting the specimen should provide the microbiologist with a working diagnosis, so that the latter may then consider the addition of special media which might help isolate those microorganisms consistent with the clinical impression. If the specimen contains large or inhibitory amounts of the antimicrobial agent administered, it may be necessary to dilute the specimen.

If the organisms suspected of being present are not frequently encountered in the particular laboratory, a representative of such microorganisms should be cultured in duplicate media to ensure that proper support and conditions for their proliferation are provided and to permit comparison of cultural characteristics when the unknown organism has grown. The use of such control cultures has been advocated by some for years. It is a helpful measure which frequently aids in a speedier recognition of organisms encountered only rarely.

As mentioned earlier, certain specimens require dilution. This is especially true of urine specimens submitted for quantitative microbiological analysis. There are many ways of ascertaining the number of bacteria per milliliter of urine. Whichever the method chosen, it must be of proven efficacy and accuracy when performed in the laboratory. The most common approach is the use of a calibrated loop that delivers a specified volume of urine to an agar plate, which is then streaked. Another approach is diluting the specimen through several tubes until a 1:100 dilution has been obtained. A small portion, usually 0.1 ml, is then delivered to the plates used for assessing the number of bacteria, and this small volume is then spread across the plates with a glass spreader, with a bacteriological loop or needle, or with the pipette itself. Finally, dilution by the shake-agar technique is used occasionally. Various dilutions of the urine specimen are added to melted, measured volumes of agar. The agar and urine are mixed, and the mixture is delivered into a petri dish where it is allowed to harden. A variety of automated, semiautomated, and screening devices are available for the detection of bacteria in urine; their advantages have recently been reviewed.

When a piece of tissue is submitted for microbiological analysis, it should be cut into small pieces or, if available, a tissue grinder should be used for maceration (see above, Collection of Specimens). If the material submitted is very viscous, it may be treated with sterile N-acetylcysteine or an equal volume of 1.5% amyl acetate, both of which have been shown not to inhibit microorganisms. This treatment is very helpful with certain body fluids and sputum.

Before primary culturing is begun, all media and equipment must be well organized on a clean bench top. A sequence of the steps to be performed should be established so that the inoculation procedure can be executed smoothly. Proper entries in workbooks, on worksheets, or on patient records should be made to indicate that each particular step has been executed. Adequate safeguards should be within easy reach; a disinfectant solution must always be on hand. The proper placement of a Bunsen burner, a microincinerator, loop carriers, or sterile swabs, if they are to be used, is a prerequisite for efficient working. Whenever possible, safety cabinets should be used. These are mandatory if aerosolization may occur during the processing of the specimen.

INOCULATION AND ISOLATION METHODOLOGY

Many methods of inoculating various media are available to the laboratory. The bacteriological inoculating loop is used most often, particularly when a homogeneous specimen is available. When broth cultures or fluid specimens are transferred to media, sterile cotton or synthetic fiber swabs, Pasteur pipettes, calibrated pipettes, syringes, and needles may be used for distributing them to various preparations of several solid and liquid media. In general, it is unwise to utilize these various transfer devices for spreading the culture across the entire agar plate. Most laboratories, therefore, plant a portion of specimens on an area near the periphery of the plate(s). After the specimen has been delivered to all of the various substrates, it is then further distributed by streaking it across the agar in a fashion that permits an ongoing dilution of the specimen. This is best accomplished with a bacteriological loop. The specimen is spread back and forth over the surface of perhaps one-quarter to one-third of the plate, with the bacteriological loop held loosely between the thumb and index finger. The weight of the bacteriological loop is allowed to exert its own pressure. There is no need to exert additional pressure with the hand. After the first quadrant has been inoculated, the plate is turned 90°, and the first few cross-streaks touch the original inoculum. After that, the laboratory worker continues to inoculate the surface of the agar, avoiding the first inoculated quadrant. When another quarter of the plate is covered, it is once more turned 90° at the end, thereby permitting a small area to be inoculated after the fourth turn has been made. This is to ensure the presence of isolated colonies. An important caveat must be stated. Familiarity with the inoculation procedure often leads to a rapid performance of this task. Personnel must contain their enthusiasm sufficiently to avoid hitting the sides of the petri dish, since such collisions create aerosols, potentially dangerous to personnel and laboratory reagents.

The microbiologist must keep in mind the purpose of the medium being inoculated. For example, highly selective media require a heavier inoculum, and one drop of the specimen may be inadequate to yield growth of the microorganism in question. Therefore, forr those media, it may be appropriate to utilize a swab to inoculate the first quadrant of the plate. Resorting to a bacteriological loop thereafter is' helpful in obtaining isolated colonies. Familiarity with the composition and purpose of each medium plus experience is the ultimate way to master this aspect of the procedure.

If the purpose of the procedure is to obtain an evenly distributed inoculum of a diluted specimen, a bent glass rod may be used. This rod is designed to make approximately a 45° angle. The side used for inoculation usually covers the radius of a standard petri dish. This sterile glass applicator is placed over the inoculum, and the plate carrying the agar is rotated until an even layer of liquid is above the agar. Care must be exercised that the inoculating volume is proper (usually 0.5 to 1.0 ml).

Animal inoculations are performed with decreasing frequency in the average clinical microbiology laboratory. It would be useful to consult standard texts describing this procedure should the need arise. As a word of caution, individuals not experienced in performing animal inoculations should send the specimens to central or reference laboratories which are equipped to provide this service. It is also important that animal facilities and care meet prescribed regulations.

Occasionally, there is a need to use the pour plate technique. Pour plates are used infrequently in the quantitation of urinary microorganisms, but the method is often used to quantitate the microorganisms in environmental cultures. For this type of inoculation, sterile pipettes are required. If dilutions are to be made, tubes containing measured amounts of sterile diluent are needed. A Bunsen burner is usually required. The other types of inoculations can all be accomplished with microincinerators, a device safer and easier to use when pathological material is involved.

If the specimen for primary inoculation is of very tenacious or friable material and is to be cultured for fungi, such as skin scrapings, nail scrapings, hairs, etc., it may be necessary to use a small sterile scalpel to cut the specimen into smaller pieces on the surface of an agar medium. This will permit the pathological specimen to be submerged into the agar and further inoculated onto the plate by using the scalpel to make additional cuts in the medium.

The inoculation methods described are to be used only for the inoculation of primary isolation media. Other methods are appropriate for transferring laboratory-grown microorganisms; these are not covered here.

MEDIA FOR PRIMARY ISOLATION

At the risk of redundancy, it must be stated once more that the purpose of using a number of different solid and broth media is twofold. It permits the isolation of many of the various microbial agents in a specimen. At the same time, a number of media select specific members of the microbial population by suppressing other microorganisms present in the specimen.

Table 1 represents the recommended media and Gram stain requirements for primary bacteriological cultures as used at the Mayo Clinic, Rochester, Minn. It is reproduced here as a guide to the selection of media as practiced in one of the outstanding institutions in the United States. It is obvious that for each medium chosen, others may be substituted. It is equally obvious that the choice of media may be expanded considerably. This recommendation encompasses the requirements for the isolation of microorganisms usually looked for in these specimens and is in no way all encompassing. Additional media must be chosen when information concerning specific microorganisms is available from the clinician. This guide is similar to others that have been published by various medium manufacturers such as BBL Microbiology Systems and Difco Laboratories or advocated in texts.

Detailed discussion of the many media for primary isolation cannot be included here. It must be remembered that the choice of which medium to use for a specific purpose represents perhaps best the mystique of clinical microbiology, a mystique compounded by habit, preferences of favorite teachers, personal experience, and occasionally solid data. The choice is also determined to a degree by the availability of commercially prepackaged agar plates and broth.

INCUBATION OF INOCULATED MEDIA

After inoculation on proper substrates, cultures should be incubated at an optimal temperature as quickly as possible. Delay in bringing certain organisms to the desired temperature may hamper their ability to grow in the desired period of time. Turbidity or other manifestations of growth in broth will be impaired. Overgrowth by undesirable microorganisms will be encouraged, especially in liquid media. The question of proper temperature has preoccupied clinical microbiologists for some time. Although it is theoretically desirable to come close to body temperature for many of the organisms involved in disease production, a number of microorganisms cannot tolerate the optimal temperature for humans. Therefore, most incubators in clinical laboratories are set at 35 °*C*, which permits the growth of virtually all organisms capable of existing at 37 °*C*. A single incubator set at 35 °*C* is obviously inadequate for many of the temperature requirements of a large clinical laboratory. Provision for changes of incubation temperature should be available and can be provided even in the smallest laboratory by the judicious use of temperature blocks, water baths, or small incubators. If only one major incubator is available, it is recommended that the temperature be set at 35 °*C*. A drawer or a closed cabinet can be used in laboratories that do not have multiple incubators for room temperature incubation. A water bath may be adjusted to a temperature beyond that of the main incubator, such as 42, 56, or 60 °*C*.

It is desirable that most cultures be inspected after 15 to 18 h of incubation, the so-called overnight incubation. Not only are most cultures readable after this time, but also it is important to the clinician and the patient that a preliminary report be issued within the first 24 h. Even a negative report on certain cultures, especially those of body fluids, has significance to the clinician. However, many of the bacteria in clinical specimens require additional incubation for visible growth. Good clinical microbiologists routinely inspect all cultures that were negative during the first 18- to 24-h period on a daily basis for a period of 1 to 2 weeks. In fact, recovery of bacteria after 4 to 5 days is not a frequent occurrence. However, this observation is offset by the isolation of bacteria after a 1- to 2-week incubation period, especially from blood cultures. Although the percentage of such delayed isolations is low, it nevertheless constitutes isolation of organisms important to the diagnosis and treatment of the patient. Only broth media containing portions of the clinical specimen are so treated. Incubation of agar media in excess of 3 days is unnecessary unless specific information for slow-growing microorganisms or a suspicion of a prolonged lag phase exists. Should the clinical information suggest that bacterial L-forms, the so-called cell wall defective variants, are suspected in the specimen submitted, hypertonic conditions should be established in the media used for primary and subculture purposes. This may be accomplished even after ordinary media have been inoculated by the addition to fluid media of suitable hypertonic solutions.

Adequate moisture is a further requirement of all incubation chambers. Even the most primitive incubator should be supplied with such a source. Many of the new incubators have water reservoirs attached or are equipped to regulate the evaporation of water into the incubator. Pans of water may be placed on the lowest shelf of an incubator not adequately equipped for this purpose. When automatic regulation of moisture is present, a relative humidity of 70 to 80% should be provided. Provision and control of atmospheric conditions are other important aspects for proper incubation for primary isolation. Many bacteria prefer a carbon dioxide atmosphere for growth. A capnophilic environment may be provided in special incubators equipped with a CO_2 gas tank which regulates adequate carbon dioxide content in the atmosphere. Burning a candle in a tightly closed jar, the candle extinction method, is still considered quite adequate in providing the increased CO_2 required by all such'bacteria. Another method of providing an increased CO_2 environment is to tip a 1% HC1 solution into a beaker containing $NaHCO_3$ previously placed in a wide-mouthed jar.

All incubators must be equipped with a thermometer. Whereas many of the newer models have mechanical recording devices, those incubators which do not should at least be monitored by a high- and lowindicator thermometer, a device which will record the lowest and highest temperature achieved between inspections. Daily checking and recording of the temperature are necessary.

Similarly, the sanitary care of walk-in incubators must be of the highest order. Accidental spills, collections of old, overlooked specimens, dust, and other soil must be avoided by meticulous attention to a cleaning schedule. Even the smaller table-top or closet-type incubators should be inspected periodically to ensure proper sanitary conditions. All personnel with access to the incubator must be aware that accidental spillage of specimens, culture media, etc., represents a serious hazard that must be quickly disinfected and cleaned up.

EXAMINATION AND IDENTIFICATION OF ISOLATES

One of the most decisive steps in clinical microbiology is the preliminary inspection of growth on primary isolation media after. the first period of incubation. Once growth has been recognized or its identity suggested by the various manifestations in and on isolation media, an accepted set of standardized procedures should be followed to initiate presumptive and definitive identification. The recognition that a potentially significant microorganism is present in a clinical specimen must result from consideration of the source of the specimen; namely, is it from a body area in which microorganisms are rarely, if ever, found in the normal state, or does the organism come from an anatomic site which has a normal microbial flora? It is important to consider not only the appearance of certain organisms on generally supportive media, but also the numbers and kind of bacteria which manifest themselves on various selective media used for the cultivation of the original specimen.

The experienced microbiologist or the technical staff require only a glance at a series of selective and general media to gain an appreciation of the organisms present. However, no one should misunderstand or underestimate the importance of this first approach to acquaint oneself with the organisms present as being sufficient for further action. It is imperative that this first scanning be followed by close inspection of each primary isolation medium employed. More often than not, such agar plates should be inspected with the aid of a hand lens, an illuminated magnifying lens, if available, or, best of all, a stereoscopic microscope. Much of the morphology of bacteria, fungi, and yeasts on primary isolation media is a self-taught pattern recognition system which combines the aspects of colonial morphology on one medium with the appearance of the same organism on various other selective media. Even the absence of detectable colonial growth on one medium is important to the identification. It is customary for laboratorians in the clinical microbiology laboratory to utilize magnification to search out and inspect beta-hemolytic streptococcal and possibly pneumococcal colonies. However, an inspection of all plates with some magnification helps to acquaint workers in the clinical microbiology laboratory with subtle differences in colonial morphology on a variety of media. Colony inspection with magnification should be encouraged in all laboratories. Even the laboratory which does not receive large numbers of clinical specimens can afford to train neophytes in clinical microbiology by providing them with known cultures which can be studied on the various media used in that laboratory. Learning to describe the morphological characteristics of each significant bacterium on each medium will enhance the ability to recognize these organisms in mixtures that are common from contaminated body areas and also make possible a prompt preliminary decision when the bacteria appear in cultures of blood and other body fluids.

The characterization of colonial morphology follows well-established guidelines. There are many authoritative texts which will acquaint clinical microbiologists with the easiest manner of describing the appearance of a colony. Many methods are useful in recording these particular differences. Some prefer to sketch the appearance of the colony in their notebooks. Others utilize terms that describe the appearance. It is also helpful to establish comparable size differences of the colonies. Although "eye-balling" and guessing the size of a colony lack all quantitative aspects, it is helpful to note these impressions. Recording the findings of colonial morphology aids in the final recognition of organisms if such data are recorded immediately and permanently, rather than attempting to reconstruct the reasons for further studies long after the original impressions of the primary colonies have become blurred in the mind of the worker.

The properties of surface colonies on solid media can be described very simply as summarized in many of the standard texts of microbiology. It is important to note whether the shape of a colony is circular, irregular, radiate, or rhizoid. The size of the colony should be expressed in millimeters. Those who use calipers or millimeter rules for the measurement of zones of inhibition in antimicrobial susceptibility tests have learned to make such measurements quickly and accurately. The elevation of the colony must be described either by sketching or by words such as raised, low convex, convex or dome shaped, umbonate, umbilicate, and with or without beveled margins. Wilson and Miles add to this description the structure of the colony, which they describe as amorphous, fine, medium, or coarsely granular, filamentous, or curved. The surface may be described as smooth, contoured, beaten copper, rough, fine, medium, or coarsely granular, ringed, papillate, dull, or glistening. The word matt has been used to describe the appearance of colonies somewhere between the dull and glistening states. The edge of the colony may be described as entire, undulate, lobate, crenated, erose, fimbriate, curved, or effuse. The pigmentation of a colony should be detected by reflected or transmitted light. Pigment may be described as fluorescent, iridescent, or opalescent, as well as by color description. It should also be noted whether these pigments are carotenoid or water insoluble, i.e., confined to the colony itself or diffused into the surrounding medium. Another variable is the opacity of the colony. Opacity may be classified as transparent, translucent, or opaque. The consistency of the colony becomes apparent to the experienced microbiologist; therefore, descriptions such as butyrous, viscid, friable, and membranous may help in characterizing the colonial morphology as well as serving as a guide to the handling of a colony should it be necessary to subculture it. The manner in which a colony behaves when it is picked for making a Gram stain may serve as a distinctive characteristic with a particular medium. This is referred to as emulsifiability and is either easy or difficult; it may be noted that the colony forms a homogeneous or granular suspension or that it remains membranous when rubbed in a drop of water with an inoculating needle. These notations are helpful, especially when training individuals to describe the appearance of colonies, and continue to aid in the recognition of organisms by their colonial morphology. Certain organisms will produce characteristic odors on various media. The ability to distinguish some bacteria by the odor produced can be of great help. However, odor detection is to be done with care and caution and is not totally reliable.

Growth in fluid media helps in the presumptive identification of microorganisms. Fluid media may be evaluated by first noting the degree of growth present. It is reported as none, scant, moderate, or abundant. Another variable is the degree of turbidity. This may be recorded as absent or, if present, as slight, moderate, or dense. Another useful descriptive observation is whether the turbidity is uniform, granular, or flocculent. Variation in these properties will occur in different fluid media. It is not possible to compare the appearance of an organism in thioglycolate broth with its appearance in brain heart infusion broth. The presence or absence of precipitates in broth media is also important. The degree of precipitate-slight, moderate, or abundant-should be noted. The precipitates may also be described as powdery, granular, flocculent, membranous, or viscid. In some instances, slight shaking will resuspend the precipitates formed, whereas no effect will be noted with others. The formation of a pellicle on the surface of the broth is still another characteristic that should be observed. If a pellicle is present, it may have a ring of growth around the inner surfaces of the tube. The pellicle may be thick, thin, smooth, granular, or rough. This manifestation may also be destroyed by shaking or may remain intact.

Growth on selective media may constitute a clue to the identification of certain microorganisms. Given Tables present summaries of some of the bacteria which may be recognized presumptively at this stage. It must be emphasized that it is not the purpose of these tables to serve as other than a guide to help the microbiologist decide whether to identify a particular organism further. This decision is influenced by the source of the specimen, the microorganism recovered, and the history of the patient. As an illustration, consider the recovery of *E. coli* from a stool specimen of a 30-year-old male. If all the colonial morphology pat-terns indicate that this organism is *E. coli* and that it is not being confused with other members of the

Enterobacteriaceae family, especially that it is not a *Shigella* species, or that such an identification error is of no consequence in this specimen, then no further work needs to be done. This is true especially if the history of the patient does not indicate any gastrointestinal symptoms. The same organism isolated from a urine specimen from the same individual would require additional tests to identify it.

The clinical microbiologist must decide which of the organisms isolated on primary culture must be identified. There are no *easy* guidelines which would cover all eventualities. If, after care-ful inspection, the organisms represent the normal microbiota of the part of the body from which the specimen was obtained, further identification of these organisms by additional testing is superfluous. If the specimen was obtained from an anatomical site which is normally sterile, the laboratory must proceed to establish the identification of genus and species, but not at the expense of presumptively reporting the initial impressions.

This need for presumptive identification is in response to the major responsibility of the clinical microbiology laboratory. As mentioned at the onset, it is the function of the clinical microbi-ology laboratory to report as accurately and quickly as possible the presence of a significant microorganism in a clinical specimen and to provide the clinician with a guide to antimicrobial therapy. This aim can be met by the judicious application of the principles outlined in this chapter and by the early and quick use of diagnostic tests available for the recognition of certain microor-ganisms. An example is the isolation and recognition of a *Staphylococcus sp.* from a specimen. A coagulase test would quickly confirm that this organism is *S. aureus* without requiring additional testing.

This does not rule out the possible future need to perform phage typing for epidemiological purposes. Similarly, the recognition of either *Salmonella spp.* or *Shigella spp.* in a stool specimen may require the use of grouping antisera to confirm the identity of the organism. The species or serotype designation may be obtained, and biochemical tests to confirm the identification of the organism may be carried out. However, the certainty and authority with which one can report the presence of such an organism in a specimen are enhanced largely by the ability to perform rapid tests leading to strong presumptive identification.

The special circumstances of the patient frequently require the testing of a microorganism further than would appear necessary at first. Also, the persistent isolation of an organism from a usually sterile body area ought to lead to identification tests, even if under ordinary circumstances such an organism would be described as nonpathogenic. Early presumptive reporting is imperative. It is the practice in many laboratories to render a verbal report to the clinician. This is followed by a written report reiterating the preliminary nature of the verbal report. Laboratory personnel (and clinicians) must also be aware that occasionally the preliminary report may prove to be incorrect. One should never hesitate to retract the initial report and indicate the correctness of the final report. This is done with explanations if required. Clinical microbiologists must not hesitate to alert the physician and the infection committee of the institution when they have presumptive evidence of organisms that are potentially dangerous to the institution and the community at large.

The classical purpose of primary isolation is the separation of microorganisms into groups which can be identified according to the principles of sound microbiology. Primary isolation constitutes the starting point for the detective work which leads to the diagnosis, if not the identification, of microorganisms in a manner expounded by the experts to guide the reader to the level of genus and species recognition. The dynamic state of the intimate human biosphere makes it imperative that the clinical microbiologist remain aware of the changes which bring new populations into this environment and that the role of these newcomers in disease production or complication be recognized with dispatch. This responsibility can be met only by the judicious use of the chapters which follow.

2

INFECTIOUS DISEASES

Varicella zoster virus (VZV) is one of the 7 known herpesviruses that affect humans. These viruses are especially well known for their unique parasite-host relationship of latency following primary infection. VZV infec-tion is first manifested as chickenpox, a very common childhood disease. The virus is believed to gain access to sensory nerve endings due to its presence in skin vesicles during varicella, and latency is established in dorsal root ganglia. Recently, the presence of VZV DNA and RNA has been demon-strated in human ganglia at autopsy by both polymerase chain reaction (PCR) and in situ hybridization.

At some time, usually many years after the primary infection, latent VZV may reactivate and cause zoster. Both varicella and zoster are usually self-limited in the normal host, although protracted post-herpetic pain may follow zoster, especially in the elderly. Either clinical condition may result in severe morbidity and even mortality in immunocompromised individuals. Because of the ability of VZV to cause latent infection, complete eradication of the virus by vaccination as has occurred with smallpox is not likely to ever be achieved. Therefore, VZV can be predicted to remain an important cause of morbidity especially in certain parts of the population, despite the availability of a vaccine and antiviral therapy.

The intricate interrelationship between VZV and the human host is probably the result of a delicate balance between the host's immune system and the virus. Changes in the immune system of the host, whether due to age, disease, chemotherapy, or radiotherapy may hamper the balance or create a situation in which the latent virus reactivates, manifesting as zoster.

IMMUNITY TO VZV

We measure protection against infection by in vitro correlates of immu-nity: antibodies, interferon production, lymphocyte transformation, and cytotoxicity assays, for instance. Surrogate markers of immunity, however, do not always indicate true protection from infection. Immunity to VZV is exceedingly complex and may be incomplete in perhaps 15% of the population, who may either develop zoster or experience a second attack of chickenpox, the latter a rather unusual event. Antibodies play a role in prevention of second attacks of varicella, acting in concert with various forms of cell-mediated immunity *(CMI)*. That *CMI* is of great importance in recovery from VZV-induced disease is illustrated by the observation that children with agammaglobulinemia recover successfully from varicella, while those deficient in *CMI* are at risk for fatal infection. Antibodies are not sufficient for prevention of zoster, as zoster occurs in patients who have antibodies to VZV. Reinfection with VZV, manifested as varicella, has been observed in antibody-positive individuals, years after natural varicella as well as after immunization. However, antibodies are not without specific effects on VZV; it has been clearly shown that modification of varicella can be successfully accomplished by passive immunization with varicella-zoster immune globulin (VZIG).

The following observations further illustrate the potential role of CMI in defense against VZV infection. In a study of immunocompromised patients, those who had severe infections had a more sluggish CMI response to VZV, measured by lymphocyte blastogenesis, than those who recovered unevent-fully. In contrast there were no differences in humoral immune responses. Poor recovery from varicella has also been linked to faulty production of interferon, and the disease has been successfully treated with this substance. Cytotoxic T lymphocyte reactions to VZV that are HLA-linked have recently been demonstrated in persons with prior chickenpox. Cyto-toxicity to VZV is mediated not only by CD8 (cytotoxic/suppressor) lympho-cytes but also by CD4 (helper) cells. Finally, antibody-dependent cellular cytotoxicity (ADCC), which may play a role in host defense against VZV, has been demonstrated in vitro.

During the period of latency, antibodies to VZV and specific CMI can be measured. Studies on aging subjects, who are at high risk to develop zoster, provide a means to explore possible immunologic perturbations preceding zoster. A relative change in subpopulations of T cells with aging, similar to those sometimes observed in autoimmune diseases, occurs. A specific decline in CMI to VZV in aging has been demonstrated. Lympho-kine production, such as the T cell growth factor interleukin-2, demonstrates age-dependent decrease in protein and RNA levels. Decrease in macro-phage function has been reported in the elderly. Presumably, these may be related to development of zoster although studies on patients before and after zoster would be extremely difficult if not impossible to perform. Prospective longitudinal studies of specific antibodies and CMI in leukemic children who have received varicella vaccine have been accomplished. They have implicated poor cellular rather than humoral immunity to VZV as a predisposing factor in zoster.

EPIDEMIOLOGY OF VZV INFECTION

A significant epidemiologic factor in varicella is related to the observa-tion that VZV spreads rather poorly in tropical and subtropical climates. Therefore, when adults from such an area travel to a country with a temperate climate, such as the United States, they have a high degree of susceptibility to varicella. This has resulted in significant development of varicella, classically a childhood disease, in adults, who often have a more severe course of illness.

One might predict that another epidemiologic change associated with im-munocompromised individuals and the elderly might be occurring. Numbers of living immunocompromised patients are ever increasing, as a result of success in chemotherapy for malignancy and prolonged immunosuppression in organ transplant recipients. Mean life expectancy has also been prolonged in healthy populations. These factors can probably be expected to contribute to a higher incidence of zoster than in the past in the developed world.

VZV IN AIDS PATIENTS

An apparently new clinical syndrome of prolonged, chronic infection with VZV has recently been described in patients with AIDS. In some cases prolonged therapy with antiviral agents has been required to control these infections, and in some cases VZV resistant to therapy has also developed. Chronic VZV lesions may be keratotic, resembling warts. It is unusual to be able to isolate VZV from them, but specific antigens are often demonstrable by immunofluorescence.

DIAGNOSIS OF VZV INFECTION

VZV can be isolated from skin vesicles but rarely from the pharynx. Virus isolation is a cumbersome procedure which demands several days of laboratory followup. Rapid diagnosis can be made today by several methods, including demonstration of VZV antigens in skin lesions with fluorescein-tagged monoclonal antibodies. VZV-specific IgM may also he demonstrated in serum as an indication of acute varicella or zoster.

Polymerase chain reaction (PCR) is a new technique that is extremely useful for demonstrating the presence of a virus that is difficult to propagate. PCR rapidly amplifies a specific sequence of DNA efficiently and has been recently used for the detection of VZV DNA in clinical specimens. PCR has been successfully used to detect VZV in vesicular lesions and throat swabs. PCR has also been used to demonstrate VZV DNA in cerebrospinal fluid and to distinguish rapidly between wild-type and vaccine-type VZV. There is no doubt that PCR will become one of the routinely used rapid diagnostic methods in the near future, after the obstacles of standard-ization and contamination have been solved.

TREATMENT OF VZV INFECTIONS

The most frequently used specific antiviral therapy for VZV infection is the drug acyclovir (ACV) which selectively inhibits viral DNA. ACV has minimal side effects, and it may be administered by mouth as well as intravenously. It is, however, poorly absorbed orally and it must be admin-istered in large doses by this route. In a recent study, oral ACV was found to have a beneficial effect on the course of varicella in healthy children compared to placebo when it was administered within 24 h of onset of rash. The drug reduced the total number of lesions and shortened the course of illness by about 1 day. One of the difficulties with such a study is that a drug effect is rather difficult to show in a disease that is normally as mild as varicella in the otherwise healthy child. Although the cost of a 5-day course of ACV for varicella is only about US $ 50 it is not usually recommended for routine use because the disease is so mild in children and the expected benefit so small. Exact indications for its use in otherwise healthy children remain to be determined. Studies of treatment of adults with varicella with oral ACV have suggested some benefit although the studies were uncontrolled.

The clinician is therefore faced with the difficulty of deciding which, if any, otherwise healthy patients to treat. One possibility is to treat secondary household cases which are known to be more severe than index cases, and to treat patients above 13 years of age. A dose of 80 mg/kg/day divided into 4 doses is given orally, with a maximum of 800 mg/dose, for 5 days. It is not usually necessary to treat the rare otherwise healthy child who develops zoster.

The main indication for the use of ACV in VZV infection is in immuno-compromised individuals, in particular those who are at great risk to develop, or who have severe varicella, and those with severe zoster. In such cases ACV should be administered intravenously, at a dose of 500 mg/m^2/ dose, three times a day, unless renal function is impaired. Immunocompro-mised patients who have not received either varicella vaccine or VZIG who develop chickenpox, should be treated intravenously, in the hospital, with ACV, as early as possible after appearance of rash. Children at high risk to develop severe varicella include those with an underlying malignancy (especially leukemia and lymphoma), with acquired or congenital immune deficiency (including AIDS), and those receiving large doses of corticoste-roids (over 1 mg/kg/day of prednisone). Fatal varicella has been reported after short-term high-dose corticosteroids for asthma [37, 38]. It is probably not necessary to treat all immunocompromised patients who develop zoster with intravenous ACV; for milder cases, oral ACV may be initiated.

VARCELLA: ACTIVE IMMUNIZATION

VZIG is well known to be useful for passive immunization of high-risk children against chickenpox. Active immunization can be accomplished with live attenuated varicella vaccine, now licensed in Japan and Korea for use in healthy and immunocompromised children, and for immunocompromised children in some European countries. In the United States, the vaccine is not licensed, although the Federal Drug Administration now permits its use on a compassionate basis for varicella-susceptible leukemic children.

Although varicella vaccine appears to have been developed mainly to protect immunocompromised children against severe varicella, it has be-come more and more accepted that healthy children have much to gain from immunization against chickenpox. The best responses to the vaccine have been observed in healthy children, with a seroconversion rate of about 95%, after one dose. Positive antibody responses are long-lasting. Adverse effects, such as rash, are mild and rare. Protection of healthy vaccinated children, upon exposure to VZV, is about 85%, with the other 15% develop-ing modified varicella. Of 3,000 children immunized in USA only 2 have developed possible zoster.

Varicella vaccine may be especially beneficial for leukemic children, although they are somewhat difficult to immunize, because about 50% who are still receiving maintenance chemotherapy develop complications such as a moderate rash and fever, for which treatment with ACV is required. Often ACV can be given orally. Two doses of vaccine are given, since only 85% seroconvert after the first dose. Certain conditions should be met for eligibility for varicella vaccine: being at least 1 year in full remission from leukemia, suspension of chemotherapy including steroids, for at least 1 week: before and 1 week after the first dose of vaccine, and a peripheral Mood lymphocyte count greater than 700/mm^3.

Vaccine type VZV had been recovered from the rash of some leukemic children, and secondary spread to about 15% of varicella susceptibles in the household, has been observed. Transmission of vaccine virus from a leuke-mic vaccinee does not occur if there is no rash. Contact infections are mild, with few skin lesions and no systemic signs, indicating that the vaccine virus is attenuated. About 85% of vaccinated leukemic children have been protected from varicella, an efficacy rate resembling protection in healthy, vaccinated children, and significantly higher than that observed in vaccinat-ed adults. About 25% of leukemics lose detectable VZV antibodies after a year; however, many re-seroconvert without symptoms after exposure.

Zoster occurs in about 15% of leukemic children who have had natural varicella. Leukemic vaccinees are therefore a sentinel population in whom to compare the incidence of zoster after immunization. According to the collaborative study and other studies of leukemic children as well, the incidence of zoster after vaccination is lower than after the natural disease.

Problems with Varicella Vaccine, and a Perspective on the Future

Some of the major problems in wide scale utilization of VZV vaccine are the difficulty in defining a susceptible population (unless the vaccine is given to every child), and the evaluation of acquisition of immunity/protection following vaccination. Sensitive and accurate serological methods in com-mercial kits, are emerging now and some of them seem quite promising. These include mainly ELISA (enzyme linked immunosorbent assay) and the recently described latex agglutination method. Unfortunately, a number of post-immunization sera that are positive bý the sensitive fluorescent antibody to membrane antigen (FAMA) assay are negative by both ELISA and latex agglutination. In addition, some falsely positive ELISA tests have been recorded.

Another problem concerns the spread of the vaccine virus to contacts. Fortunately, the changes of spread are directly related to the number of skin lesions, so that spread is unusual, especially in healthy children who are unlikely to develop a vaccine-associated rash. Moreover, due to the attenuation of the virus, the danger from such infection is remote.

Healthy children are already being routinely immunized against varicel-la in the Orient, and it is predicted that this vaccine will be licensed in the US in the near future. VZV vaccine can be administered in combination with measles, mumps and rubella (MMR).

3

VIRAL INFECTIONS

Timely diagnosis of a specific virus infection can affect patient care through the use of antivirals and/or immune therapies, by avoiding the use of unwarranted antibiotic therapy, or through cohorting to control nosocomial transmission. Rapid tests are available to complement or substitute for isolation in cell culture and serology so that virtually all of the common viral pathogens can be identified rapidly enough to affect patient management.

METHODS AND STRATEGIES OF DIAGNOSTIC VIROLOGY

Identification of a specific virus begins with the clinical diagnosis. With the diagnosis as part of any scheme, there are two general methodological approaches: (1) The isolation of virus in cell cultures, and (2) the detection of specific viral components (antigens, nucleic acids, or enzymes) directly in the specimen. Comprehensive laboratories offer virus isolation and rapid detection.

Isolation depends on viral replication in a susceptible cell culture. The procedure is often too time consuming to impact on patient care. Nevertheless, isolation remains the standard of sensitivity and specificity for viral diagnosis. Virus isolation also offers the advantage of identifying the unsuspected virus and multiple viral infections. A study addressing this issue reported that 27% of the specimens negative for respiratory syncytial virus (RSV) by a rapid test yielded a virus other than RSV in cell cultures. A retrospective study of 6 years' experience reported dual infections in 8% of specimens that were also reactive for RSV. The non-RSV identifications and the dual infections would have been missed if isolation in cell cultures was omitted.

SPECIMEN COLLECTION AND TRANSPORT

The diagnosis of specific viruses is dependent on the quality of the specimen received in the laboratory. In general, the quantity of virus is greatest early in the illness and decreases as resolution of disease occurs. Infectious virus is best preserved in the patient. It is preferable to delay collection a few hours if the delay shortens the interval between obtaining the specimen from the patient and processing it in the laboratory.

Virus Isolation

Most specimens should be placed in a transport medium; commercial transport media are available but a buffered bacteriological broth will suffice. Urines and spinal fluids are exceptions and may be held and transported in sterile containers. Specimens should be held at 4 °C and transported cold whenever possible; freezing should be avoided unless the specimen cannot be processed for 48 h or more.

Direct Detection of Viral Components

The nature of the test will dictate the collection and transport procedures, particularly if commercial tests are used. In general, the guidelines for virus isolation should be followed.

TECHNIQUES APPLIED TO RAPID DIAGNOSIS

Considerably more viral protein and nucleic acid are made in an infected cell than are incorporated into virions; indeed, only I in 10-1,000 virions may be infectious. It is assumed that this is also true in an infected patient. Theoretically, detection of these viral components should provide a more sensitive and rapid basis for specific diagnosis than isolation. There are, however, no methods routinely available that are universally accepted to be more sensitive or specific than isolation in cell culture. The following are brief descriptions of the rapid techniques that are most frequently used.

IMMUNOFLUORESCENCE

Immunofluorescence (IF) is the most widely used technique for rapid testing. Intact cells are obtained from the patient and stained with specific antibody labelled with fluorescein-isothiocyanate (FITC) in a direct or indirect format. The specimen is viewed with a microscope that produces a transmission band in the range of 400-500 nm. Reactive cells appear bright green.

An important feature of IF is that the quality of the specimen may be assessed since the number and condition of the cells can be evaluated during viewing. No other rapid procedure includes this feature. Objectivity blends with subjectivity, however, as time passes the intensity of the fluorescence diminishes and the patterns become less distinct. It is imperative that skilled and experienced technologists perform the test. The added need for expensive instrumentation requires that IF be done in comprehensive laboratories.

Time-Resolved Fluorescence Immunoassay (TRFI)

TRFI utilizes a fluorescent conjugate with a much larger Stokes shift than FITC to give a longer time to measure emission of fluorescence. The procedure increases sensitivity but requires specialized instrumentation and has not been extensively utilized.

ENZYME IMMUNOASSAY

Enzyme immunoassay (EIA) is the rapid method used most frequently after IF and is a common commercial format. The test construction routinely consists of a solid phase, usually plastic, to which a specific viral antibody is attached. Viral antigen in the specimen is 'captured' by the antibody. The remainder of the specimen is removed by washing and the immobilized antigen is detected by a second specific antibody which has an attached enzyme. Following the addition of a substrate, the enzyme and the complexed viral antigen are detected by a color change. EIA reactions have objective endpoints and may be quantitated by using a spectrophotometer. Automated tests can examine large numbers of specimens and the data analyzed by computer. The common configurations detect between 1,000 and 10,000 infectious particles.

Enzyme Immunofiltration (EIf)

EIF assays capture and bind antigen nonspecifically to membranes in individual cassettes; an EIA reaction identifies the presence of specific antigen. Attractive features of the commercially available EIF tests are internal reference standards to detect invalid results. EIF has produced rapid (less than 30 min) and convenient tests for areas where laboratory facilities may be minimal or non-existent.

Immunoenzymatic Staining

In a format analogous to that described for IF, the detection label is an enzyme rather than FITC. An enzymatic reaction with a chromogenic substrate produces a precipitate that is viewed with a light microscope. The technique is not used extensively for rapid testing but has found general acceptance in anatomical pathology.

DETECTION OF VIRAL NUCLEIC ACID

The nature of the genetic code, the high avidity of complementary strands of nucleic acid, and the rapid kinetics of association provide a basis for tests with exquisite specificity and high sensitivity.

Hybridization

Complementary sequences to unique portions of a viral genome are selected as probes for those segments. The genomic sequences of most clinically important viruses are known and computer programs are available to assist in the selection of complementary probe sequences. Probes are manufactured using DNA synthesizers or by encoding restriction enzyme fragments in plasmids. Test constructions, components, and terms (dot blot, slot blot, southern and northern blots) may differ. Essentially, viral DNA is liberated from the clinical specimen and denatured into single strands. The separated strands are reacted with the complementary strand, the probe. The development of nonisotopically labeled probes has increased the usefulness of the technique. Common labels include those used in EIA tests.

Hybridization is most often applied to diagnostic situations where other methods or reagents are not available. Sensitivity is usually not greater than that of other rapid techniques or culture and reactions may take 1-2 days to perform. Important specialized applications have derived. In situ hybridization provides a diagnosis of virus in tissue sections including formalin-fixed, paraffin-embedded sections. Automation and direct colorimetric detection of reactions permits large numbers of sections to be tested in hours.

Polymerase Chain Reaction (PCR)

In contrast to hybridization, the detection of nucleic acid by the PCR is not limited by the amount of target viral DNA. The PCR exponentially amplifies a target sequence of DNA so that millions of copies of the segment may be available for detection.

A unique sequence of the viral genome is selected as the target segment. Short pieces of DNA (primers) that are complementary to the terminal sequences of the opposite strands of the target segment are synthesized. DNA is extracted from specimens and placed with the primers in a reaction mixture. The reaction proceeds in thermal cycles where for periods of minutes reaction temperatures are programmed to denature the DNA into single strands, allow the primers to anneal to the target segment, and permit DNA synthesis. The DNA product is identified by conventional methods. Detection of RNA genomes requires synthesis of a cDNA template using reverse transcriptase before amplification can be initiated.

The PCR is applicable to the diagnosis of all viruses and relevant reports are in the literature. In several instances PCR is the method of choice. The widespread use of PCR is restricted, however, by the current licensing constraints.

ELECTRON MICROSCOPY

The lack of sensitivity, expense of instrumentation, and requirement for specialized training has discouraged the general utilization of electron microscopy (EM). Additionally, virion morphology is usually not specific at the level of genus or species upon which clinical decisions are based.

SHELL VIAL TECHNIQUE

The shell vial technique (SVT) merges the specificity and sensitivity of isolation in cell culture with immune detection into a rapid test. Originally reported for cytomegalovirus (CMV), the method has been applied to most viruses that replicate in cell cultures and for which monoclonal antibodies are available. A cylindrical container (the shell vial) containing a cell culture growing on a circular cover slip is inoculated with a specimen. Following centrifugation at low speed, and incubation for 24 h or less, the coverslip is prepared for detection of antigen. CMV is detected by an antibody to the immediate early antigen, the first protein detectable following productive infection. Similarly, other viruses are detected by antibodies to the earliest antigens expressed or to antigens expressed in large quantity.

SEROLOGY

All serology is retrospective since the production and subsequent detection of antibody follows exposure to antigen. In lieu of virus detection, schemes that can evaluate an active infection by discerning specific antibodies in an acute serum are most valuable to patient management; examples are serology for the hepatitis viruses and Epstein-Barr virus (EBV).

IF is commonly used to detect antibodies, particularly IgM, but has inherent problems. Endpoints are difficult to determine and nonspecific fluorescence is a chronic problem. Results should be interpreted cautiously particularly for serology of human herpes viruses. IgG should be extracted from sera to be tested for IgM antibody to remove blocking antibodies, and substrate for rheumatoid factor. A useful but complicated modification is the 'anti-complement immunofluorescent test'.

EIA is becoming the standard test method for viral serology. The 'antibody-capture EIA' is a modification that circumvents many of the problems associated with the traditional format and does not require prior treatment of test sera.

DIAGNOSIS OF SPECIFIC VIRUSES

The clinical diagnosis points to the suspected virus(es). A comprehensive laboratory contributes to the differential diagnosis and on fewer occasions focuses on one suspected virus. In the following section the most commonly used methods for identification of specific viruses will be reviewed in terms of differential diagnosis and in some instances individual viruses.

RESPIRATORY VIRUSES

Many viruses are associated with respiratory disease. The seriousness of the infections and the availability of antivirals, however, place influenza A and RSV in the forefront of diagnostic efforts.

The typical specimens are a nasopharyngeal (NP) wash (preferred), a NP swab, or a throat swab and/or gargle. The NP wash provides material for virus isolation, cells for direct IF, and is usually satisfactory for EIA; throat swabs are not recommended if RSV or the parainfluenza viruses (PIV) are suspected. Cell cultures that permit isolation of influenza A or B, RSV, PIV 1, 2, and 3, adenoviruses, rhinoviruses, and enteroviruses are utilized.

Direct IF can be performed on the same specimen for influenza A or B, PIV 1, 2, and 3, adenovirus, and RSV; in our experience, the IF test for adenovirus is not as sensitive as it is for the other viruses. The shell vial test has been applied to these viruses but there are few published reports to judge efficacy.

Commercial EIA tests are available for RSV. The tests are convenient, require 1.5-3 h to perform, but are less sensitive than IF or isolation.

The EIF formats available as rapid tests for RSV and influenza A require approximately 20 min. They are satisfactory as screening tests, particularly during peak periods of transmission. Results should be confirmed, however, by more reliable procedures.

HUMAN HERPES VIRUSES

The natural histories and pathogenesis of these viruses requires that each virus be given individual attention.

Herpes Simplex Virus (HSV)

HSV replicates in several different cell cultures although sensitivities vary. In the best of conditions, 50% *of* HSV isolates may be identified by cytopathic effect (CPE) in 24 h and 90% within 3 days. Use of the SVT can increase identification to >90% in 24 h or less. HSV,1 may be distinguished from HSV-2 with monoclonal antibodies.

Direct IF of cells taken from the base of a vesicular lesion is an effective rapid technique and is particularly valuable in discerning HSV from varicella-zoster virus (VZV). The traditional Tzanck preparation is often difficult to interpret, is not as sensitive as direct IF, and does not differentiate between HSV and VZV.

The best EIA tests available for HSV are not reliable in detecting HSV in asymptomatic excreters. Sensitivities and specificities may be acceptable for symptomatic patients, particularly when specimens can be obtained directly from lesions.

Isolation of virus from a brain biopsy remains the only definitive diagnosis of HSV encephalitis. Less invasive procedures usually are not sufficiently sensitive until the second week of illness. The PCR has been used in a small number of patients to detect HSV DNA in spinal fluids and may prove to be beneficial for rapid diagnosis.

Serology can distinguish primary infection by seroconversion but is unreliable in recurrent infections.

Varicella-Zoster Virus (VZV)

Isolation of VZV is difficult and can only be accomplished from the clear fluid of vesicular lesions. The SVT enhances isolation rates and accelerates the time to detection of virus. Direct IF of cells obtained from the base of a vesicular lesion is the most useful test. In contrast to isolation, the lesion may be in virtually any stage of evolution and still yield suitable material for examination. In combination with the analogous procedure for HSV, a definitive diagnosis of VZV versus HSV may be made in less than an hour.

Serological tests are used to assess immunity to VZV. Results obtained with the fluorescent antibody membrane antigen (FAMA) tests correlate well with resistance or susceptibility to infection. FAMA is the standard test, but the anti-complement immunofluorescent (ACIF) test gives comparable results and is easier to perform. Commercially available EIA tests are commonly used but are less sensitive than FAMA or ACIF.

Cytomegalovirus

Cytomegalovirus is readily isolated from clinical material but the time required (an average of 7-12 days) compromises the usefulness of this approach. The SVT has provided a useful rapid method but requires a comprehensive laboratory with an experienced staff. Examination of polymorphonuclear leukocytes by IF using a monoclonal antibody to IE antigen is a rapid method for assessing viremia and may have prognostic value.

Epstein-Barr Virus

EBV may be isolated in B cells obtained from human umbilical cord blood. The procedure is laborious, requires 2-4 weeks, and is not part of routine diagnosis.

Approximately 85-90% of all patients mainly above 4 years of age with infectious mononucleosis (IM) have heterophil antibodies which persist for about 6 months. Patients with primary infections but without clinical IM may not have heterophil antibodies while other diseases may give false positive results. A specific diagnosis or an evaluation of EBV status is accomplished by looking for antibodies to 3 different antigen complexes: (1) virus capsid antigen (VCA); (2) early antigen (EA) with diffuse (D) or restricted (R) patterns, and (3) Epstein-Barr nuclear antigen (EBNA). IgG and IgM antibodies to VCA appear during primary infection; the IgG antibodies persist for life while the IgM antibodies disappear after several months. Patients with nasopharyngeal carcinoma have high titers of IgA antibody to VCA. IgG antibodies to EA also appear during acute infection but are usually undetectable 6 months later. Most laboratories do not discern between R and D patterns in routine diagnosis. Importantly, antibodies to EA can reappear during reactivations of EBV. IgG antibodies to EBNA are detectable weeks to months after infection and persist for life. Current investigations into antibody responses to the 6 individual components of EBNA may provide in the future prognostic information relative to EBV diseases. A subject is considered immune if anti-IgG VCA and EBNA antibodies are present.

VIRAL HEPATITIS

Diagnosis is based upon serology and in some instances detection of antigen. Tests for 4 hepatitis viruses are available.

Hepatitis A Virus (HAV)

IgM antibodies are detectable between 4 and 6 weeks after infection followed thereafter by IgG antibodies. The IgM antibodies are usually undetectable by 12 to 16 weeks; IgG antibody persists for life.

Hepatitis B Virus (HBV)

Hepatitis B surface antigen (HBsAg) and hepatitis B e antigen (HBeAg) are detectable prior to the appearance of symptoms and during the acute phase. Antibody to hepatitis B core antigen (HBcAg) appears during the antigenemia phase but is not indicative of immunity. Antibody to HBsAg (anti-HBs) is associated with immunity and appears when HBsAg is no longer detectable. Patients without anti-HBs but with antibody to core antigen are potentially infectious; persistence of antigen is indicative of chronic infection.

Hepatitis C Virus (HCV)

HCV is one of the causes of non-A non-B (NANB) hepatitis. An impressive feat of molecular biology resulted in a recombinant antigen of HCV which has been adapted to a serological test. Screening is, therefore, currently available and further elucidation of HCV and its natural history are in progress.

Hepatitis D Virus (HDV)

Replication of HDV, formerly termed the delta agent, requires simultaneous HBV infection. Detection of antigen in liver by immunohistochemical techniques is the only reliable way to diagnose virus replication although a hybridization procedure was reported. A commercial EIA for antibody to HDV is the most commonly used test for diagnosis.

VIRAL GASTROENTERITIS

Rotavirus, adenoviruses 40 and 41, Norwalk viruses, astroviruses, calciviruses, and coronaviruses are commonly grouped as viruses causing gastroenteritis (GE). None of the GI viruses replicates readily in cell cultures and some of them have never been successfully propagated. Each of the viruses has a distinctive morphology and may be diagnosed by EM. Adenoviruses 40 and 41, however, cannot be discerned from other non-GE adenoviruses that may also be present in stool specimens. EIA tests are available commercially for rotavirus and adenoviruses 40 and 41; latex agglutination (LA) tests are also available for rotavirus. LA tests are simple to perform. rapid (30 min), and highly specific.

HUMAN IMMUNODEFICIENCY VIRUS

There are two types of HIV, HIV-1 and HIV-2. Diagnostic procedun continue to focus on HIV-1 but are likely to combine analysis for both virusi in the near future. Tests for antibody, antigen, and the technology to isola virus are available. The following descriptions were derived from inform; tion relative to HIV-1.

Detection ofAntibody

Antibody to HIV is generally screened for by El tests although an indirect IF antibody assay is also available. A reacti'% serum must be confirmed by western blot analysis, a procedure that dete mines antibody-reactive bands of specific viral proteins. The best accepte criteria for confirmation is the presence of any two of the p24, gp4 I , an gp 120/160 bands. There is an indefinite period between infection an the detection of antibody. Antibodies become detectable to most of the HI proteins. Exacerbation of disease is usually associated with diminution i antibody to the core protein.

Detection of Antigen. Following infection and before antibody is detectable there is antigenemia characterized by detection of p24 antigen in serum i plasma; as antibody to p24 becomes detectable, the quantity of antige diminishes to undetectable levels. Antigenemia is detectable again with tl advent of disease. EIA tests are most commonly used to detect p24. IF tes and hybridization procedures are also available to detect HIV in cells but a less frequently employed. HIV may also be isolated in culture from peripher blood lymphocytes. Isolation is laborious and lacks sensitivity, particularly patients whose CD4 cells are not severely depleted. The PCR provides tl most sensitive method for detecting HIV including proviral DNA and vir RNA and hence active or latent infection. The method is particularly valuab in situations where other procedures are not usable or are insensitive.

PAPILLOMAVIRUSES

The papillomaviruses have not been grown in cell culture and immui reagents are not available. Beyond the clinical diagnosis of warts, laboratory diagnosis is based upon the detection of nucleic acid. The many types papillomaviruses can also be discerned in the same manner.

PARVOVIRUS

Only one strain of human parvovirus is recognized, B19. The virus, which is the cause of erythema infectiosum (fifth disease), may precipitate aplastic crisis in patients with chronic hemolytic anemia, and has been associated with hydrops fetalis. Laboratory diagnosis is based upon detection of IgM antibody and/or detection of nucleic acid. The procedures are not routinely available and specimens usually must be sent to a reference laboratory.

PROSPECTIVE

The development of antivirals should continue to propel the evolution of rapid diagnostic methods. Formats currently used are likely to be improved but there are limitations. Coincident with these advances, one can anticipate an increase in resistance to antivirals and the need to make rapid determinations of antiviral sensitivity of isolates. Currently, that is accomplished almost exclusively by infectivity assays using cell culture. New and rapid procedures will have to be developed. The future may belong to versatile, yet exquisitely sensitive and specific procedures like the PCR that can be adapted to many diagnostic problems.

Questions regarding latency, reactivation of latent viruses, and pathogenesis need to be addressed from a diagnostic viewpoint. The identification of virulent genes or genes responsible for pathogenesis and their transcription products may provide the diagnostic strategy for the next generation of tests for viral diagnosis.

4

Antiviral Therapy and Prophylaxis

Antivirai therapy has three main mechanisms of action: (1) inhibition of viral cellular attachment; (2) inhibition of viral intracellular replication; (3) enhancement of the host immune response through 'passive' immunization (i.e. interferon, intravenous immunoglobulin). Most viruses produce irreversible cellular damage early in the course of the disease and pathogenicity is often augmented by the host immune response. Unfortunately, neither of these phenomena is likely to be influenced by antiviral compounds, unless given early in the disease course.

Moreover, evaluation of antiviral drugs (table 1) can be difficult since most viral infections are self-resolving, and carefully designed placebocontrolled studies with well-defined endpoints are necessary. For this reason, the clearest results are often those studies investigating life-threatening viral infections such as herpes simplex virus (HSV) or human immunodeficiency virus (HIV). This chapter will focus on common viral diseases in pediatrics and currently available treatment modalities, especially noting more recent concepts in pathogenesis and therapy.

HERPES SIMPLEX VIRUS INFECTIONS

Two drugs have proven effective in treating systemic herpes

TABLE 4.1. ANTIVIRAL DRUGS

Drug	Indication	Route	Dosage	Interval
Acyclovir	genital HSV	P.O.	30 mg/kg/day	t.i.d.
	HSV encephalitis	i.v.	30 mg/kg/day	q8°
	HSV in immuno-compromised host	P.O.	30 mg/kg/day	q4°
	neonatal HSV	i.v.	30 mg/kg/day	q4°
	VZV	P.O.	80 mg/kg/day	q.i.d.
	VZV in immuno-compromised host	i.v.	50 mg/kg/day	q8°
Amantidine	**Influenza A**	**P.O.**	**5-9 mg/kg/day**	**b.i.d.-t.i.d.**
Didanosine (ddl)	**HIV**	**P.O.**	**1.1-1.4 m², 100**	**mg b.i.d.**
			0.8-1 m², 75 mg	
			0.5-0.7 m², 50 mg	
			<0.4 m²,25 mg	
Foscarnet	acyclovir-resistant HSV	i.v.	120 mg/kg/day	q8°
	CMV retinitis	i.v.	180 mg/kg/day	q8°
Ganciclovir	CMV in immuno-compromised host	i.v. initially	10 mg/kg/day	q 12°
		mainte-nance	5 mg/kg/day	q 24°
Idoxuridine	HSV kerato-conjunctivitis	topical		q 1-2°
Ribavirin	RSV infection	inhala-tion by SPAG over 12-18°	6 g	q 24°
Rimantidine	Influenza A	P.O.	5 mg/kg/day max 200 mg/day	b.i.d.
Trifluridine	HSV kerato-conjunctivitis	topical		q2°
Vidaribine	neonatal HSV	i.v.	10-30 mg/kg/day	q24°
	HSV kerato-conjunctivitis	topical		q 3°
Vitamin A	measles requiring hospitalization	P.O.	100,000 IU	x 1
Zidovudine	HIV infection	P.O.	180 mg/m²/dose	q4-6°

SPAG - Small particle aerosol generator.

simplex infections. The first, vidarabine, acts as multiple sites in intracellular viral replication to inhibit both viral RNA and DNA synthesis. However, this drug has serious clinical and pharmacological disadvantages, including central nervous system (CNS) toxicity, poor absorption from the gastrointestinal (GI) tract and poor water solubility. It must be administered intravenously in large fluid volumes making administration to young infants problematic in the usual setting of CNS infection with increased intracranial pressure.

Hence, vidarabine has been largely supplanted by acyclovir, an analogue of the nucleoside thymidine, that inhibits herpes virus DNA synthesis. Its activity is dependent on phosphorylation by a specific viral enzyme, thymidine kinase (TK). Although human cells produce the same enzyme, only the viral enzyme will phosphorylate the drug efficiently, leading to therapeutic levels in infected cells and low toxicity. Viral resistance to acyclovir has been described, especially among immunosuppressed patients infected with HIV, due to a mutation in the gene encoding TK. This mutation has subsequently been associated with treatment -failures and prolonged disease. Because CSI absorption is only 20-30%, treatment of potentially lifethreatening infection should be accomplished intravenously. Two other systemic drugs that exhibit against HSV are ganciclovir and trisodium phosphonoformate (foscarnet).

NEONATAL HERPES SIMPLEX

The incidence of neonatal herpes, a systemic disease that leads to early multi-organ failure and death, is low (1 in 8,000 live births). However, it is one of the most devastating viral infections. Peripartum vertical transmission occurs at a much higher rate (33%) in infants born to mothers with primary genital herpes infection as compared to those born to mothers with recurrent herpes infection (3%), presumably due to the transfer of maternal antibodies and/or lower viral inocula. Acyclovir is the drug of choice, although vidarabine appears clinically as effective. Although there are insufficient data on the benefits of prophylactic therapy given to infants born to mothers with active disease, therapy with systemic preparations should be considered, especially in infants with other risk factors such as prematurity or prolonged rupture of membranes.

HERPES SIMPLEX ENCEPHALITIS

Acyclovir is the drug of choice. A large US multicenter study has shown it to be superior to vidarabine; mortality was 54% in the vidarabine group, as compared to 28% in the acyclovir group. In addition, at 6 months only 14% of vidarabine-treated patients were functioning normally as compared to 38% of acyclovir-treated patients. Acyclovir should, therefore, be started empirically in any child with focal encephalitis in whom herpetic encephalitis is a prime diagnostic consideration. Because of the clinical safety of acyclovir, therapy should not be delayed while further diagnostic procedures confirm or refute the diagnosis of HSV infection.

MUCOCUTANEOUS HERPES SIMPLEX

When gingivostomatitis is mild in nature, specific therapy is not needed. However, patients with severe disease should be treated. An oral suspension is available and is well tolerated in young children. Herpetic whitlow is another common complication of herpes occasionally seen in children with herpes labialis or gingivostomatitis who suck their fingers. If infection is extensive, antiviral therapy should be given. Recurrent infections may be less frequent in response to prophylactic therapy.

GENITAL HERPES SIMPLEX

Treatment of infected mothers with genital HSV may decrease the risk of congenital infection and its devastating consequences. A recent, preliminary study of late third trimester acyclovir administration demonstrates it to be safe and efficacious in preventing recurrences. To date, no congenital malformations have been noted in over 110 first trimester exposures to acyclovir.

Acyclovir treatment of primary genital HSV infection decreases the frequency and severity of recurrences. Several studies of daily acyclovir prophylaxis have convincingly demonstrated its long-term efficacy and safety in preventing recurrences. Treatment for 3 years with 400 mg acyclovir orally twice daily in patients with frequently recurrent genital herpes simplex (> 12/year) resulted in a mean recurrence rate of 1/year during the third year with 61 % of the patients recurrence-free for that year.

HERPES INFECTIONS IN THE IMMUNOSUPPRESSED CHILD

Orofacial herpes simplex infection is very common; 20-40% of young adults have recurrent orofacial herpes infection and primary infection usually occurs in early childhood. It is for this reason that children with primary or secondary immunosuppression are at high risk for activation and subsequent dissemination. Acyclovir is effective in treating such recurrences and is also useful when given prophylactically before and after bone marrow transplantation. However, we favor the approach of treating recurrences aggressively as intermittent subtherapeutic doses of oral acyclovir (which may occur in patients poorly compliant with prophylaxis) have been associated with emergence of resistance herpes simplex strains.

Acyclovir-resistant strains of herpes simplex are a special problem among patients with HIV infection in whom recurrent lesions often cannot be controlled adequately. A recent study showed the superiority of foscarnet over vidarabine in acyclovir-resistant mucocutaneous herpes simplex in patients with the acquired immunodeficiency syndrome (AIDS).

HERPETIC KERATOCONJUNCTIVITIS

Ocular infections can be readily treated with topical agents, some of which are available only by this route, because of systemic toxicity. Idoxuridine, vidarabine and trifluridine are the most commonly used agents, but acyclovir ointment is also available and can be used as an alternative. The need for systemic therapy is controversial. We recommend systemic therapy for ocular HSV infection in infants below the age of 12 months and children with diffuse cutaneous herpetic infection. Because herpes keratitis may precede encephalitis, particularly when topical therapy is given, careful monitoring of the CNS is advised. Isolates of herpes simplex obtained from ocular infections can show varying sensitivities. Careful monitoring of clinical progression and, if infection persists, early consideration of resistance with appropriate adjustments it treatment is the recommended approach.

EPSTEIN-BARR VIRUS INFECTION

Epstein-Barr virus (EBV), unlike herpes simplex, does not induce TK production in infected cells and, therefore, minimal amounts of acyclovir are phosphorylated. However, the specific viral DNA polymerase is very sensi-. tive to small amounts of acyclovir and there is a pronounced in vitro antiviral effect. A placebo-controlled study in patients with infectious mononucleosis found no major clinical advantage of acyclovir over placebo, although *viral shedding was reduced in the treatment group. We recommend* acyclov therapy for EBV infections only in immunocompromised patients. *There at* reports of regression of oral hairy leukoplakia (associated with EBV) wit] acyclovir and this drug may ameliorate the EBV-associated lymphoprolifera tive symptom. We further recommend treatment of immunocompe tent patients with severe manifestations of EBV infectious mononucleosis such as encephalitis. Additional controlled clinical data in support of each of these 'indications' would be desirable.

VARICELLA ZOSTER VIRUS (VZV) INFECTION

Varicella in the normal infant or child is usually a mild self-limited disease. Two recent double-61W placebo-controlled trials of oral acyclovir initiated within 24 h of onset ofchickenpox showed that the treatment group defervesced *sooner, had fewer lesions with earlier cutaneous* healing (2 vs. 3 days) but did not have a different rate of complications. Acyclovir should not be used routinely; however, treatment may be justified in certain situations. Although not specifically addressed by this work, early initiation of acyclovir therapy is likely to be of benefit in secondary household cases, adolescents, very young children or others who may have more severe illness. Acyclovir usage in other instances for this benign disease will remain an issue of cost-effectiveness.

VZV infections can be extremely severe in immunocompromised patients and the mortality rate among untreated patients is high. Acyclovir is recommended at maximal dose (up to 50 mg/kg/day i.v.) as VZV sensitivity to this compound is significantly lower than herpes simplex.

CYTOMEGALOVIRUS INFECTION

Cytomegalovirus (CMV) is very prevalent among humans and seroconversion is apparent in 20-100% of the adult population. Ganciclovir is much more potent than acyclovir in inhibiting CMV replication but can cause significant bone marrow toxicity. Both drugs have been used in the prophylaxis of CMV disease in immunocompromised patients and ganciclovir in treating acute life-threatening infections.

Specific high titer anti-CMV immunoglobulin has also been used as an adjunct to treat life-threatening CMV infections and has recently been licensed in the U.S. Currently, the most effective regimen for treatment of CMV in the immunocompromised host appears to be concomitant therapy wth IVIG and ganciclovir. Either nonspecific IVIG or CMV-specific IVIG can be used as both have proven effective.

CONGENITAL CYTOMEGALOVIRUS INFECTION

Congenital CMV occurs frequently as a result of primary maternal infection during pregnancy (1% of all pregnancies in the US). Unfortunately, prevention of the disease by effective vaccines or by early detection and therapy with drugs that can be used safely during pregnancy is not yet possible. Acyclovir given to infants with congenital CMV yielded no clinical improvement. Ongoing studies are evaluating ganciclovir effects on congenital CMV. At this point, because of the potential morbidity and mortality associated with congenital CMV, we recommend treatment with ganciclovir only in infants with disseminated life-threatening disease with careful monitoring for toxicity, such as bone marrow suppression.

MEASLES

Although widespread immunization programs are common, measles outbreaks remain a problem in both developed and third-world countries. A safe, proven therapeutic modality for young infants and immunocompromised patients still does not exist. However, ribavirin, a synthetic nucleoside is active in vitro against measles and has been used clinically. Case reports in immunodeficient patients indicate that combined aerosol and intravenous administration of ribavirin may decrease mortality, although other reports showed no benefit.

Additionally, a correlation exists between measles case fatality rates and vitamin A deficiency. In fact, vitamin A-treated children, with measles demonstrated decreased morbidity and mortality. We recommend that children with severe measles be given vitamin A supplements and treated with ribavirin.

INFLUENZA

Influenza outbreaks cause significant mortality among the elderly, as well as considerable morbidity in infants and children with immunodeficiency, bronchopulmonary dysplasia, cystic fibrosis and other chronic illnesses. Obviously, immunization is the preferred strategy. Amantadine and its analogue, rimantadine, if given within 48 h of infection can reduce disease severity in influenza A infection. Prophylactic therapy of unimmunized patients may be beneficial during influenza A outbreaks and can prevent or eliminate symptoms among children with chronic diseases. A large multicenter study using aerosolized ribavirin, a broad spectrum antiviral drug, showed that treated patients defervesced 16 h sooner but had no significant improvement in respiratory symptoms or other vital signs.

RESPIRATORY SYNCYTIAL VIRUS (RSV)

Several studies have documented the efficacy of ribavirin in children with RSV infections, including a recent study among intubated infants. The cost effectiveness of ribavirin in healthy infants with RSV infection is controversial.

We recommend ribavirin treatment of any compromised infant with RSV infection and of previously healthy children with severe manifestations of RSV, such as impending respiratory failure or very young age (below the age of 2 months).

HUMAN IMMUNODEFICIENCY VIRUS INFECTION

It is beyond the scope of this chapter to completely address therapy of HIV infection. The most useful drug currently is zidovudine which has been studied intensively in adults and has been shown to ameliorate the disease process and prolong survival. Zidovudine is a reverse transcriptase inhibitor, and bone marrow toxicity is the primary concern with long-term administration. Other drugs such as dideoxyinosine, dideoxycytosine and foscarnet are also being studied. Combination therapies appear to have synergistic antiviral activity and may avoid the toxicities associated with high-dose single-drug therapy. In a recent report of a phase I study of the combination of zidovudine and dideoxyinosine, there was enhanced antiviral activity with this combination as evidenced by increased CD4 counts and decreased p24 antigen titers.

RHINOVIRUSES

Rhinoviruses are the usual cause of the 'common cold' syndrome. Their antigenic variability (more than 100 serotypes) has prevented the development of an efficacious vaccine, so far. Intranasal interferons have been used as prophylaxis, but side effects were considerable. Other experimental approaches include the use of monoclonal antibodies or specific viral attachment or replication inhibitors.

VIRAL HEPATITIS

Viral hepatitis affects at least 400,000 people annually in the United States, most often causing a debilitating, though self-limited illness. A minority suffer progressive hepatic dysfunction and death. The five specific classifications include hepatitis A, B, delta, C, and E. Non-A non-B hepatitis, thought to be the principal cause of transfusion hepatitis today, represents a heterogenous group of at least two distinct viral agents, including the recently discovered hepatitis C virus.

Hepatitis A is an RNA virus which is transmitted by the oral-fecal route. Outbreaks of hepatitis A have been reported in daycare centers with children from 1-2 years being the most commonly affected. Current postexposure prophylaxis consists of immune globulin 0.02 ml/kg i.m. given within 2 weeks of exposure. Ribavirin and amantadine have activity in vitro against the hepatitis A virus, although clinical studies of efficacy are lacking.

Hepatitis B is less prevalent among children but vertical transmission from HBsAg-positive mothers poses a major preventive problem. Current recommendations are to administer hepatitis B immune globulin (HBIG) 0.5 ml intramuscularly to infants born to HBsAg-positive mothers as soon as possible after birth followed by three doses (0, 1 and 6 months) of the recombinant vaccine. The first must be administered within 7 days of the HBIG with a different syringe at a separate site. Universal immunization with the recombinant vaccine starting in infancy has recently been adopted in the US.

Hepatitis B infection and transfusion-derived hepatitis C are diseases with major acute morbidities (liver failure) and chronic sequelae (chronicactive hepatitis, cirrhosis and hepatocellular carcinoma). Treatment modalities include interferon and ribavirin. Several subtypes of interferon have shown promise in treating acute non-A, non-B/C hepatitis and preventing progression to chronic non-A, non-B/C hepatitis. Patients who were treated with interferon had lower transaminase levels at 6, 12 and 24 months posttreatment and had no detectable antibody to hepatitis C virus. However, only 20% of patients with chronic hepatitis remain in remission after treatment. Further studies are necessary to delineate the role for, and preferred doses of, interferon in the treatment of non-A, non-B/C hepatitis.

HUMAN PAPILLOMA VIRUSES

The most common manifestation of papilloma viruses are genital, hand and plantar warts. Papilloma viruses also cause juvenile laryngeal papillomatosis, a condition that may lead to life-threatening respiratory compromise in affected infants. Thsese lesions can be extremely difficult to completely eradicate surgically. Therapy is limited to tissue destruction by keratolytic agents, laser ablation or freezing. Intralesional administration of interferon is an investigational method that has resulted in frequent recurrent rates. Recently, combined CO_2-laser surgery and recombinant interferon-alpha (IFN-α) have been found to be more effective than surgery alone.

5

NEONATAL INFECTIOUS DISEASES

Various etiologic agents including viruses, bacteria, fungi and protozi cause infections in the newborn period, which may be acquired at differe times during pregnancy, at or before delivery, or from the external enviro ment. This chapter focuses mainly on neonatal bacterial septicemia, wi emphasis on predisposing factors, laboratory evaluation and antimicrobi therapy.

Neonatal septicemia is defined as a generalized bacterial infection in the first 28 days of life documented with a positive blood culture. During the period, infection remains a significant cause of mortality and long-ter morbidity. In the preantibiotic era, mortality from neonatal sepsis exceed 90% but with the advent of improved neonatal care and antibiotics, this has been reduced to 15-75% depending on the causative organism an underlying risk factors. Several epidemiologic investigations have clear demonstrated a change in the pathogens responsible for neonatal seps Before the 1940s, the predominant organisms were group *A Streptococc* which were replaced by coliforms in the mid-40s. In the 1950s, *Staphyi coccus aureus* assumed a rather important role but gram-negative bacil particularly *Escherichia coli* predominated in the 1960s. From the 1970s the present, group *B Streptococcus (GBS)* has been responsible for most cas of neonatal sepsis, but *E. coli* is still common.

EPIDEMIOLOGY

Bacterial sepsis in the neonate occurs in 1-10 per 1,000 live births with a significantly higher rate among preterm than term neonates. With systemic bacterial infection, mortality ranges between 15 and 75% depending on the gestational age, time of occurrence of infection in the postnatal period and the availability of a neonatal intensive care facility. The immunological status of the host and other complicating conditions such as prematurity, respiratory distress syndrome, intrauterine growth retardation and pregnancy induced hypertension also influence the mortality rate in neonates.

Most, if not all, bacterial infections of the newborn are opportunistic in a sense that the infecting organism may be readily isolated from the cervicovagina'. canal, skin and the gastrointestinal (GI) tract of normal individuals. According to information from the Centers for Disease Control and various other reference centers, the predominant pathogens causing neonatal sepsis and meningitis in the United States in order of decreasing frequency are GBS, *E. coli, Listeria monocytogenes* and *Enterococci*. Coagulase-negative *Staphylococcus* and *Pseudomonas* nosocomial pathogens are mostly seen in premature and sick infants needing prolonged intensive care. Slight variations in the causative agents may be observed among different nurseries, which emphasize the importance of local surveillance of bacterial flora.

ROUTES OF INFECTION

Prior to rupture of the fetal membranes, the infant's environment is usually sterile. However, certain organisms gain entry and reach the fetus causing a localized or generalized infection. Major routes of infection are either vertical (transplacental, ascending, intrapartum) or horizontal (community acquired and nosocomial). Through maternal infection, the bacteria crosses the fetoplacental barrier causing transplacental infection as exemplified by congenital syphilis, *L. monocytogenes* and *Mycobacterium tuberculosis*. An alternate route is by an ascending infection through the cervicovaginal canal with either intact or ruptured membranes. Typically, this is seen in early-onset GBS sepsis and herpes simplex type II infection. Intrapartum acquisition of infection as the infant descends through the birth canal is another mode and is seen in cases of gonorrhea ophthalmia, *E. coli* sepsis and chlamydia pneumonia and conjunctivitis. This may also be the mechanism involved in late-onset GBS sepsis, in which the infant is colonized and disease-free at birth but later becomes symptomatic and develops sepsis or, more commonly, men ingitis. Intrapartum colonization and subsequent neonatal infection alsc occurs when fetal and maternal blood mix as may occur with hepatitis B. cytomegalovirus (CMV) and human immunodeficiency virus (HIV). Lastly, the infective organism may be introduced to the infant after birth in the nursery or at home.

TABLE 5.1. ROUTES OF NEONATAL INFECTION

A. Vertical transmission

1. Transplacental: maternal infection crosses the fetoplacental barrier, e.g. congenital syphilis, *L. monocytogenes, ML tuberculosis, CMV, HIV,* toxoplasmosis, rubella, hepatitis B
2. Ascending: from cervicovaginal canal affecting intact or ruptured membranes, e.g. GBS (early onset), HSV type 11
3. Intrapartum: as the infant descends birth canal, e.g. gonorrhea ophthalmia, *E. coli,* GBS (late onset), *Chaamydia,* HSV type II Fetal and maternal blood mixtures, e.g. CMV, hepatitis B, HIV

B. Horizontal transmission

1. Community acquired: from home environment, e.g. H. *influenzae, S. pneumoniae, N.-meningitidis*
2. Nosocomial infections: in the sick newborn that requires prolonged care in intensive care unit, *e.g. Staphylococcus, Candida sp., CMV, Pseudomonas sp.*

HOST FACTORS

Prematurity is the single most important risk factor for neonatal sepsis. Inherent to the preterm newborn are deficiencies of the immune system such as decreased immunoglobulin synthesis, specific antibody response, delayed switch from IgM to other immunoglobulin isotypes, decreased or absent pathogen-specific IgG, such as antibodies against GBS and other IgG sub-classes, impaired opsonic functions, deformability, decreased white blood cells chemotactic and phagocytic capabilities. Adequate transplacental transfer of IgG does not occur until after 32 weeks of gestation, therefore, these premature infants are hypogammaglobulinemic at birth. Christensen et. al., Haque et. al. and Von Muralt and Sidiropoulos observed a dramatic reduction in mortality associated with neonatal sepsis after intravenous gamma globulin infusion, further supporting the association of hypogammaglobulinemia and increased risk for sepsis noted in premature infants.

Other host factors that may increase the risk of neonatal sepsis include a 5-min Apgar score <6, twin gestation, male sex, congenital skin abnormalities causing breakdown of anatomical barriers and other factors that expose the infant to opportunistic organisms. These include endotracheal intubation, chest tube insertion and umbilical catheterization.

MATERNAL FACTORS

Maternal historical information is of utmost importance in assessing fetal risk of infection. Infants born to mothers with clinical amnionitis (maternal fever, uterine tenderness, fetal tachycardia and purulent amniotic fluid) are at increased risk. Genital colonization with pathogenic bacteria, i.e. GBS, has an incidence risk of 1% for neonatal sepsis. This increases 4-fold if accompanied by prolonged rupture of membranes (PROM) greater than 18 h, or maternal fever. The degree of colonization is also directly proportional to the rate of transmission.

Other reported maternal risk factors believed to affect the incidence and outcome of neonatal sepsis include maternal urinary tract infection (presumably by increasing the risk for preterm labor), coitus prior to delivery, internal monitoring devices and unsterile obstetrical examinations (i.e. digital) post-rupture of membranes.

PATHOGEN'S VIRULENCE

The pathogenesis of infection is determined by the interaction between the host and bacteria. Bacterial characteristics that favor multiplication and subsequent infection include virulence and invasiveness, ability to produce toxins (i.e. exotoxins and gram-negative endotoxins), slime *(Staphylococci),* antiphagocytic factors, enzyme production, intracellular pathogenicity and antigenic heterogeneity.

Certain bacteria are more invasive than others because of adherence factors such as fimbriae in some strains of *Streptococci* or *E. coli* pili. The release of endotoxin seen with gram-negative bacteria causes hypotension, leukopenia, activation of C3, complement cascade and disseminated int vascular coagulopathy. Examples of enzyme production include coagul; from *S. aureus,* streptolysin 0 from group *A Streptococcus* and leucocic from certain *Staphylococcus* species. Certain capsular (K) antigenic types *E. coli* may also be markers of virulence.

CLINICAL MANIFESTATIONS

The septic neonate presents with nonspecific signs and symptoms such poor feeding, lethargy or irritability that are usually first observed by the mother or nurse. Both early as well as late infections may present with the same symptom complex. It is not uncommon to find respiratory distress, abdominal distention, jaundice, mottling of the skin, irregular breathing, apnea, hepatosplenomegaly, hypotonia and lethargy on physical examination. Similar nonspecific signs may also be seen in several other noninfectious conditions such as respiratory distress syndrome, intracranial hemorhage, hematologic, metabolic, GI, cardiovascular and endocrine disorders. A high index of suspicion, therefore, plays a crucial role in the evaluation and early treatment of a neonate especially with one or more of the risk factors previously mentioned.

GROUP B BETA-HEMOLYTIC STREPTOCOCCUS

In recent years, GBS has attracted increasing attention as a major treatable cause of devastating neonatal infection. Investigators have focused on maternal GBS carriage and occurrence of neonatal infection. [Studies have shown that maternal colonization rate at the time of delivery ranges between 20 and 30%. Half of the infants born to these mothers are colonized. The incidence of GBS disease is estimated to be 1 for every 100 colonized infants. Infant colonization and disease are affected by the degree of maternal colonization. Ferrieri looked at the acquisition rate of babies born to mothers from whom GBS colonization was quantitated into light to heavy colonization. She observed that the transmission rates were 67% for infants born to mothers with heavy, 43% for those born to mothers with moderate and 31% for infants of mothers with light GBS colonization in the vagina. The rate of positive blood culture was also directly proportional to the degree of colonization.

EARLY-ONSET GBS DISEASE

Early-onset disease is defined as a systemic illness and bacteremia during the first week of life. This form of infection is highly associated with obstetrical complications such as PROM, fetal distress and prematurity. The incidence is 3.7/1,000 live births for early onset and 1.7/1,000 live births for late-onset disease. Baker and Edwards described early-onset infection as a result of an ascending infection occurring immediately'before delivery. In a 10-year review of neonatal sepsis (1979-1988) at Yale New Haven Hospital, early-onset infection was dominated by GBS (55%) followed by *E. coli* (14%), all other sepsis (10%), *Haemophilus inf uenzae* (8%) and gramnegative anaerobes (4%). The clinical symptoms are nonspecific and represent primary sepsis with subsequent involvement of virtually every tissue site. Mortality rates range between 50 and 75% and the serotypes responsible for early GBS disease are equally distributed among la, Ib, Ic, II and III. The most prominent clinical feature is pneumonia. Bacteremia and septic shock are frequent and meningitis occurs in approximately 30% of the cases.

LATE-ONSET GBS DISEASE

Late-onset sepsis occurs in infants after the first week of life. These infants usually have an unremarkable birth and perinatal history. In contrast to earlyonset disease, late-onset disease tends to be focal often affecting the central nervous system (CNS) causing meningitis. The mechanism of transmission of the pathogen remains unclear. It is speculated that these infants acquire the infection by harboring the organisms in mucosal sites. They then develop transient asymptomatic bacteremia seeding metastatic areas such as bones, joints and meninges. The mortality rate is lower in late-onset disease (14-48%). GBS serotype III is responsible for 90% of cases of late-onset GBS infection.

E. COLI

E. coli belongs to the family *Enterobacteriaciae* and accounts for 70% of all gram-negative neonatal infections. *E. coli* and GBS constitute 5075% of all cases of neonatal bacterial sepsis. There are over *100 E. coli* strains that cause disease and approximately 80% of all *E. coli* meningitis and 40% of septicemia are caused by a specific capsular antigen, K1. It has been shown in animal models that the presence of K1 antigen is highly associated with lethality and in human infection, the outcome is correlated with the presence, association and the persistence of K 1 antigen. This may be related to interference with opsonization. *E. coli* is the second most common pathogen in both early (14%) and late (31%) onset disease.

Epidemiologic studies of this organism show that as much as 50% of all pregnant women are colonized and so are 20-30% of all newborn infants. About 60% of all infants born to K 1 antigen-positive mothers are colonized at birth; however, one out of 100 colonized infants (like GBS) develops infection. The clinical features of *E. coli* sepsis are in general similar to those infections caused by other organisms.

LISTERIA MONOCYTOGENES

Listeria monocytogenes is a gram-negative intracellular bacillus that is responsible for sporadic and epidemic infections in humans; serotypes IVb, Ib and Ia being responsible for most cases reported.

In most instances, neonatal infection is caused by transplacental route although ascending infection via cervicovaginal route has been described. Human infection is transmitted as food-borne caused by improper pasteurization of dairy products or vegetables contaminated with sheep manure containing *Listeria*.

Classically, maternal infection presents with fever, abdominal pain premature contractions, leukocytosis, non-productive cough, pharyngitis headaches, rigors, decreased fetal movements and meconium stained amniotic fluid, which represents meconium liquefaction by microorganisms. Intrauterine infection characteristically produces rash, hepatosplenomegaly and pneumonia. Histologically, miliary microabscesses are found teeming with polymorphonuclear leukocytes and bacteria. Diagnosis is made by gram staining of amniotic fluid obtained through amniocentesis. Disease in the newborn is divided into two groups defined by age: early- and late-onset disease.

EARLY-ONSET LISTERIOSIS

Early-onset disease is thought to be secondary to either maternal bacteremia infecting the fetus transplacentally as early as the 19th week of gestation or during passage through the vaginal canal. The infant presents prematurely with respiratory distress due to aspiration of contaminated amniotic fluid, meconium staining in spite of prematurity, dark skin papules or petechiae and hepatosplenomegaly. The mortality rate reported by most centers is approximately 50%. It is associated with high-morbidity secondary to the illeffects of prolonged mechanical ventilation.

LATE-ONSET LISTERIOSIS

Late-onset disease occurs following colonization of the infant during passage via the birth canal. The infant develops signs and symptoms of infection at 1-8 weeks of life. As with late-onset GBS, infection frequently involves the CNS. Meningitis is seen in 95% of all cases of late-onset disease. The cerebrospinal fluid (CSF) in meningitis caused by *Listeria* characteristically shows hypoglycorrhachia and pleocytosis often with predominance of lymphocytes. Gram staining may show the bacilli. Mortality due to late-onset disease is generally low and long-term morbidity uncommon.

COAGULASE-NEGATIVE STAPHYLOCOCCI

During the last decade, advancements in neonatal care has permitted the survival of very low birth weight (VLBW, i.e. birth weight < 1,500 g) infants. Infections due to coagulase-negative

Staphylococci are currently the most common nosocomial infection in VLBW infants, those chronically hospitalized and infants with central venous catheters, endotracheal tubes, chest tubes and ventriculo-peritoneal (VP) shunts. Mortality associated with these organisms ranges between 0-15%.

Coagulase-negative Staphylococci are ubiquitous organisms found on the skin and mucous membranes that are acquired early in life either by passage through the birth canal or exposure to caretakers. All preterm neonates are colonized by the 4th day of life. Infection may be introduced by a break in the skin, or through the GI or respiratory tract.

Factors responsible for the increased predilection of this type of infection among premature infants include deficient opsonic activity which is important in the clearance of bacteria, frequent use of intravascular catheters, invasive procedures such as endotracheal intubation and chest tube insertion, slime elaboration of the bacteria which facilitates catheter colonization, adherence and antimicrobial inhibition. Clinical symptomatology is similar to the nonspecific symptomatology described previously but tends to be more indolent. It may present with septicemia, meningitis, endocarditis, necrotizing enterocolitis (NEC), pneumonia, omphalitis and abscesses.

HAEMOPHILUS INFLUENZAE

H. influenzae has recently been implicated in neonatal sepsis. Friesen and Cho in their review noted that 80% of *H. influenzae* sepsis in the neonates was caused by non-type b in contrast to >95% of type b predominating in infancy. Like sepsis caused by other organisms, *H. influenzae* is associated with obstetrical complications including PROM, genitourinary infections and prematurity. It presents as early-onset sepsis in 83% of cases. The overall mortality is 55% but is much higher (90%) among premature infants <30 weeks gestation.

CANDIDIASIS

Although systemic candidiasis is not of bacterial etiology, we feel obligated to mention it here since it is emerging as one of the most important pathogens in late-onset sepsis affecting VLBW infants. The increasing incidence is caused by the increased survival of these infants. It affects 3-4% of all VLBW infants. The risk factors associated with systemic candidiasis include prematurity, use of broad-spectrum antibiotics, duration of antimicrobial therapy, use of indwelling arterial and venous catheters, hyperalimentation, use of intralipids, prolonged endotracheal intubation and use tracheostomies.

Candida albican.F accounts for 75% of all fungal infections in tl neonate. Other species include *C. tropicalis, C. lusitania, C. parapsilosis* and *C glabrata*. The clinical course is indolent with intermittent symptoms. The insidious symptoms and the absence of a screening test may delay the diagnosis. The mortality of disseminated candidiasis may be as high as 54-60%. Recently, detection of *Candida* enolase has yielded promising results for the early diagnosis of disseminated candidiasis.

Disseminated fungal infection presents with deterioration of respiratory status, guiac positive stools, abdominal distention, temperature instability carbohydrate intolerance, candiduria, rash, thrombocytopenia and hypotension. With a high index of suspicion, diagnostic evaluation is indicated. Evaluation should include indirect retinoscopy for endophthalmitis, renal, liver and cardiac ultrasound, lumbar puncture and suprapubic aspiration of urine as candidemia may result in seeding of these organs.

DIAGNOSIS

In order to diagnose neonatal sepsis, one must carefully evaluate the infant's symptomatology, changes in vital signs and physical examination findings. Maternal and perinatal history are key in the evaluation. Isolation of bacteria from blood, urine or CSF is the gold standard and the most specific method of diagnosis. As many as 98% of all positive blood cultures will yield bacterial growth by 72 h, hence, a culture should not be labeled negative until such time. It is, however, disturbing to note that a large proportion of neonates (as many as 32/39 in one series) proven septic by postmortem examination had negative blood cultures. The blood' culture 'gold standard' is further tainted by the emergence of contaminants which further compound the decision to stop therapy. The study by St. Geme et al. suggested that enumeration of colony-forming units in a blood culture would help distinguish the neonate with bacterial sepsis from one whose blood culture had been contaminated. Furthermore, it is important to note that a blood culture taken from an infant whose mother received intrapartum antibiotics will generally be negative.

Due to the significant number of infants reported to have meningitis but negative blood cultures, a lumbar puncture should be performed on all newborns with suspected bacterial sepsis as a primary diagnosis. Sarff et al. observed that CSF white cell counts were normal in the range of 0-32 for term and 0-29 cells/mm' for preterm neonates. Normal protein values range between 20-170 and 65-150 mg/dl for term and preterm neonates, respectively, and glucose between 32-119 mg/dl in term and 24-63 mg/dl in preterm neonates.

Because a large fraction of neonatal bacterial sepsis is caused by GBS, commercially available immunologic methods of screening for this bacteria are routinely used in most nurseries. This can be applied to serum, CSF or more commonly, concentrated urine. It has a high sensitivity and specificity. Interpretation of positive urine antigen screen but negative blood culture may be cumbersome. This discrepancy may be due to a false-positive reaction secondary to cross-reacting antigens, partially treated GBS sepsis, low-grade bacteremia with insufficient blood volume sampled, GBS colonization of GI tract and skin surface contamination of urine collected in bags. Asymptomatic infants with positive latex agglutination warrant close observation but infants with positive cultures deserve treatment.

A helpful but nonspecific laboratory test used to identify sepsis is the complete blood count. Normal white blood cell (WBC) counts in neonates range between 5,000 and 25,000 cells/mm^3 and more than a third of neonates with proven sepsis will have a normal WBC count. Manroe et al. investigated neutrophil indices, i.e. total neutrophil count, immature neutrophil count and immature/total neutrophil ratio, and found a 100% negative predictive value for bacterial infection if all these three indices were normal. Several conflicting data have been reported regarding interpretation and predictive value of these indices; therefore, other authors have modified these criteria. Amidst these discrepancies, neutropenia was noted to be the single most accurate predictor of bacterial infection.

The use of other nonspecific adjunctive tests have been evaluated. These include C-reactive protein (CRP), platelet count, erythrocyte sedimentation rate (ESR), plasma fibronectin level and other acute-phase reactants such as haptoglobin and orosomucoid. These tests are nonspecific and, taken singly, are of little clinical significance. CRP may be elevated in noninfectious conditions such as meconium aspiration syndrome, asphyxia, shock and premature rupture of membranes. A decreasing CRP level is helpful in following treatment response. ESR is elevated in inflammatory conditions as well as in Coombs-positive hemolytic disease. It may be normal in DIC due to consumption of fibrinogen which in turn decreases rouleaux formation. Again, this is a helpful simple bedside procedure that helps in assessing treatment response.

Plasma fibronectin is a glycoprotein that is present at low

concentrations in newborn infants. Conditions such as sepsis, prematurity, RDS, asphyxia and malnutrition further depress the level.

MANAGEMENT

The major challenge clinicians encounter is the early and accurate identification of infected infants. This is often difficult as the symptoms are nonspecific and seen in a number of noninfectious conditions. It is important to keep in mind that the absence of risk factors should not prevent one from treating a symptomatic infant. Remember that the burden of proof lies on the clinician to prove the existence of infection, so that a symptomatic infant should be considered septic unless proven otherwise. The decision to treat is based upon a logical and rational evaluation of the infant's clinical presentation, clinical course, risk factors and laboratory information, after which one should be able to determine which of those cases initially identified represent the true infection and therefore require a full course of antibiotics and which do not.

Appropriate therapy for bacterial sepsis is twofold. The first consists of supportive measures other than antimicrobials. After establishing IV access and hydration, careful monitoring of vital signs should be initiated. Oxygen and instruments necessary for intubation and stabilization should be readily available.

Treatment should be delayed until appropriate cultures are obtained. A combination of ampicillin and an aminoglycoside is recommended. This therapy covers GBS, *Listeria, Enterococci* and most gram-negative organisms. Penicillin or ampicillin may be used with equal efficacy. Once the bacteria is isolated, the antibiotics should be continued based on sensitivities.

We advocate the initial addition of gentamicin for GBS sepsis because in vitro studies have demonstrated synergistic activity against some strains of GBS. Once MIC/MBC results ascertain that the organism is sensitive and not tolerant (MBC not 4 dilutions greater than MIC), penicillin may be continued as a single therapy. Bacterial sepsis is treated for 7-10 days while meningitis should be treated for 14 days (table 2). Large inoculum size ($> 10^6$) and relatively resistant organisms (MIC to penicillin >0.12 and < 1) have been implicated in therapeutic failures or recurrence of GBS sepsis.

Gram-negative sepsis is treated with the same antibiotic combination for 10-14 days. In cases of meningitis, however, a third-generation cephalosporin, i.e. cefotaxime, with better blood-brain barrier penetration, may be used as a single agent or in combination with an aminoglycoside. Lumbar puncture should be repeated 24-28 h after initiation of antimicrobials and treatment continued for 21 days or 14 days following a negative culture, whichever is longer. Listeriosis should be treated with an ampicillin plus gentamicin combination for 14 days, whereas meningitis requires at least 21 days. Cephalosporins should not be used due to uniformly resistant sensitivity patterns.

TABLE 5.2. TREATMENT OF NEONATAL SEPSIS

Onset of infection	Treatment	Duration
A. Early Onset	I Empiric therapy	
	ampicillin + gentamiein	
	2 Tailored therapy	
	Bacteremia without focus	7 days
	Dacteremia with focus (i.e. lungs, urine)	10-14 days
	Meningitis with gram-negative pathogen	21 days or 2 weeks after negative culture
	gram-positive pathogen	14 days
B. Late onset	I Empiric therapy	
Community-acquired	ampicillin + cefotaxime	
Hospital-acquired	ampicillin + gentamicin + vancomycin OR vancomycin + cefotaxime	
	2 Tailored therapy	as in A

The initial empiric therapeutic regimen for late-onset sepsis is different for the infant coming from the home environment compared to the hospitalized patient who is at risk for nosocomial infection. A combination of ampicillin and a third-generation cephalosporin is indicated for those patients coming from home, in contrast to the hospitalized patients who should be started on vancomycin and a third-generation cephalosporin such as cefotaxime or ceftazidime if *Pseudomonas aeruginosa is* suspected. The presence of intravascular devices, VP shunts and NEC enhance the indication for vancomycin use. Once the pathogen is identified and antimicrobial susceptibilities known, the most appropriate antibiotics should be selected. Coagulase-negative *Staphylococci* are rarely resistant to vancomycin. Treatment should be continued for 10-14 days for uncomplicated sepsis and 14 days for uncomplicated NEC and meningitis. It may be necessary to remove the central line or VP shunt if antibiotic therapy is ineffective in eradicating the infection.

Mortality due to disseminated candidiasis is significantly reduced by prompt removal of indwelling vascular catheters and initiation of antifungal therapy. Treatment consists of amphotericin B 0.5 mg/kg/day to increase to 1.0 mg/kg/day for a total dose of 20-30 mg/kg. 5-Fluocytosine (5-FC) 100 mg/kg/day in 4 divided doses may be added to the regimen when Seeding of organs (kidneys, brain, heart) is documented or in the absence of clinical improvement. 5-FC should not be used alone due to frequent emergence of resistance. Renal and hepatic functions should be closely monitored despite the paucity of toxicities reported in the literature.

PREVENTION AND FUTURE TRENDS

Several treatment and prevention strategies for neonatal sepsis are currently under investigation. In the case of GBS sepsis, chemoprophylaxis and immunoprophylaxis have undergone several clinical trials. Antepartum chemoprophylaxis may be beneficial in full-term newborns, however, since 30% of infants with early-onset GBS disease are preterm, maternal treatment from early third or late second trimester seems unreasonable. Intrapartum chemoprophylaxis, consisting of maternal IV ampicillin given selectively is acceptable. Since only 1 % of exposed infants develop GBS disease, it is unwise to treat 100% of colonized women. Through epidemiologic studies identifying high-risk neonates, Boyer et al. noted an increased risk of GBS disease in infants whose birth weight is <2.5 kg, with PROM > 18 h and maternal intrapartum fever. This approach has the potential of eliminating >50% of early-onset GBS disease and 75% of associated deaths, diminishing the incidence of postpartum fever, minimizing drug reactions and increasing cost effectiveness. Success in postnatal preventive therapy has been inconsistent.

Transplacental transport of IgG does not begin until 28-32 weeks of gestation. Mothers often lack specific antibodies against the major neonatal pathogens, e.g. GBS, E. *coli,* or this may be of the IgM variety which do not cross the placenta. In light of these facts, IVIg administration is currently undergoing clinical investigations. Haque et al. administered IVIg prophylactically to preterm and low birth weight infants to prevent neonatal sepsis. The infection rate in the nontreated control was significantly higher (8/50) compared to the treated (2/50) group. No adverse effects of treatment were noted during the study and at 6 months follow-up.

Other investigations using prophylactic IVIg showed conflicting results. Baker noted a significant reduction in proven bacterial infections only in infants with birth weight < 1,500 g compared to the control group. There was also a reduction in the incidence of NEC. IVIg has been shown to be most effective in the prevention of GBS disease but no protection has been documented with other pathogens. In another study, 235 premature infants were enrolled in a multicenter, double-blind study using IVIg in the prevention of infection. In spite of maintaining > 500 mg/dl plasma IgG level with gamma-globulin infusions, no significant difference was observed in the incidence of documented and presumptive infections compared to placebo.

There has been limited clinical trials in the treatment of newborn bacterial sepsis with IVIg. The actual therapeutic protocols have been complicated because many interdependent and independent variables contribute to the outcome. Although the results look promising, no definite conclusions can be drawn. More recently, human monoclonal and hyperimmune polyclonal antibody containing increased concentrations of specific antibodies have been used. Findings suggest that, compared to the conventional IVIg, the human monoclonal antibody containing specific IgM and hyperimmune IVIg had better protective activity (50-90%) against types I, II, III GBS strains in doses as low as 4-20 mg/kg. The opsonic activity also significantly exceeded conventional IVIg. At this time, this is still under investigation.

Fibronectin administration has been proposed as an adjunctive immunotherapy in neonatal sepsis. There is increasing evidence suggesting its role in reticuloendothelial enhancement, improvement of phagocytic function microbial killing and opsonization, chemotaxis, phagocytic migration and adhesion. It is available and has the potential for mass production. However, prior to its clinical application, further investigations regarding metabolism, synthesis and side effects are necessary.

Christensen et al. and Cairo et al. reported dramatic improvement in the survival of septic neonates after granulocyte infusion. However, in a randomized prospective trial, Bayley et al. showed no demonstrable benefit to critically ill, ventilated infants. The major limitations of these trials stem from the inconsistency or variations in granulocyte extract preparation, process of randomization and the small sample size making statistical inference difficult.

Another new and exciting field of investigation is the use of growth factors. Recently, animal studies have shown dramatic improvement in the survival of rats infected with GBS in rats antenatally and postnatally treated with granulocyte and granulocyte-macrophage colony-stimulating factors. Again, these are still under investigation and further confirmatory studies are needed prior to clinical trials.

Lastly, the area of maternal immunization has been revived since success was seen with its use in smallpox, rubella and tetanus. Inadvertent immunization of pregnant mothers with meningococcal vaccines A and C proved protective to the infants for 3 months. Likewise, maternal *H. influenzae* immunization increased the infant's antibody levels 20-fold compared to controls. Although these results are encouraging, the strategy of maternal vaccination may not be feasible until maternal and neonatal safety are ensured.

6

PSEUDOMONAS

Of the numerous species of pseudomonads, until recently only *Pseudomonas aeruginosa, Pseudomonas pseudomallei,* and *Pseudomonas mallei* were considered important human pathogens. Since the early 1960s, however, other pseudomonads such as *Pseudomonas maltophilia* and *Pseudomonas cepacia* have been associated with infections, primarily as opportunistic pathogens of the compromised host. Prompt diagnosis of *Pseudomonas* infections, identification of the clinical isolate, and determination of specific therapy (because of the unpredictable resistance to antimicrobial agents) are necessary for a favorable prognosis.

A system is presented for the isolation and progressive identification of named and unnamed *Pseudomonas* species listed in Table 1 and their differentiation from other glucose-nonfermenting gram-negative rods encountered in clinical specimens. The current classification is based on the internal division of *Pseudomonas* into various RNA homology groups which represent natural genetic arrangements. Within the RNA groups are subgeneric groups, e.g., the fluorescent group, which contain species that share many common phenotypic properties.

IDENTIFICATION OF PSEUDOMONAS SPECIES

Isolation Media

Pseudomonads grow easily on routine primary isolation media used in bacteriology and are readily isolated from clinical specimens and from hospital environments. They are usually isolated on peptone agar media with or without infusion containing 5% sheep or rabbit blood, although media enriched with blood are not essential for their isolation. In addition to blood agar, one of the less inhibitory selectivedifferential media, e.g., Leifson deoxycholate, MacConkey, or eosin-methylene blue agar medium, should be used. Although the less inhibitory selective-differential media may increase the probability of recovering pseudomonads from specimens containing numerous other bacterial species, they do not support the growth of all pseudomonads. Pseudomonads are less frequently isolated on the more inhibitory selective differential media, e.g., Leifson deoxycholate citrate or salmonella-shigella agar. Media such as cetrimide agar, pseudomonas agar F, Pseudosel agar, and pseudomonas isolation agar containing cetyltrimethylammonium bromide (cetrimide), 2,4,4-trichloro2-hydroxydiphenyl ether (irgasan), 9chloro-9-(4-diethylaminophenyl)-10-phenylacridan *(C-390),* or similar compounds are used for the selective isolation of *P. aeruginosa* and other *Pseudomonas* species.

Identification Methods

Definitive taxonomic and biochemical clarification of glucosenon-fermenting gram-negative rods allows for easy recognition of most species by simple routine diagnostic tests. The battery of media and tests has been modified over a number of years and may need further revision as additional glucose-nonfermenting gram-negative rods are recognized.

Screening Procedure

A screening procedure designed to separate similar genera and to distinguish species within the genus *Pseudomonas* should collect the following information: odor, pigmentation, and colonial morphology; Gram reaction, somatic shape, and spore formation; motility and flagellar anatomy; mode of glucose utilization; production of hydrogen sulfide, arginine dihydrolase, and indophenol oxidase; growth at 42°C; and oxidation of glucose, xylose, lactose, and maltose.

Somatic Morphology

Pseudomonas species are straight or slightly curved rods. *Acinetobacter* species are usually very short plump rods often approaching coccoid shape, and *Moraxella* species are usually very short plump rods, predominantly in pairs and short chains and occasionally pleomorphic. *Achromobacter* species are noticeably barred when stains

TABLE 6.1: *PSEUDOMONAS* SPECIES ENCOUNTERED IN CLINICAL SPECIMENS

Name	*Synonyms*	*Name*	*Synonyms*
RNA group I		RNA group III	
Fluorescent group		Acidovorans group	
Pseudomonas aeruginosa	Bacterium aeruginosum, Bacillus pyocyaneus, Pseudomonas polycolor	Pseudomonas acidovorans	
		Pseudomonas testosteroni	
Pseudomonas fluorescens	Pseudomonas chlororaphis, Pseudomonas aureofaciens, Pseudomonas le *monnieri*	Pseudomonas delafieldii	
		RNA group IV	
Pseudomonas putida	*Pseudomonas ovalis*	Diminuta group	Group Ia
	Stutzeri group	*Pseudomonas diminuta*	*Corynebacterium vesiculare*
Pseudomonas stutzeri	Bacillus denitrificans H, *Bacterium stutzeri, Pseudomonas stanieri,* group Vb-1, group Vb-3	Pseudomonas vesicularis	
		RNA group V	
		Pseudomonas maltophilia	*Xanthomonas maltophilia, Pseudomonas melanogena,*
Pseudomonas mendocina	Group Vb-2	group I	
		Species of uncertain affiliation	
	Alcaligenes group		

Table 6.1 contd.

Table 6.1 contd.

Pseudomonas pseudoal *caligenes*	Pseudomonas alcaligenes Pseudomonas alcaligenes biovar B	Pseudomonas mesophil- ica	Bacillus extorquens, Vibrio extorquens, Pseudomonas *extorquens, Methylobacte rium extorquens, Proto* monas extorquens
Pseudomonas sp. group I	*Pseudomonas denitrificans*		
RNA group II		*Pseudomonas paucimo bilis*	Flavobacterium devorans, Flavobacterium capsula- *tum,* group IIk-1
Pseudomallei group			
Pseudomonas mallei lus mallei	Bacillus mallei, Actinobacil- Pseudomonas pertucino		
Pseudomonas pseudo mallei	Bacillus pseudomallei, Mal leomyces pseudomallei	gena	
Pseudomonas cepacia	Pseudomonas multivorans, Pseudomonas kingae, group EO-1	Pseudomonas putrefa- ciens	Achromobacter putrefaciens, *Pseudomonas rubescens, Alteromonas putrefaciens,* group Ib-1, group Ib-2
Pseudomonas gladioli	*Pseudomonas marginata, Pseudomonas alliicola*	*Pseudomonas sp.* group Ve-1	
Pseudomonas pickettii	*Pseudomonas thomasii, Pseudomonas pseudoalca*	*Pseudomonas sp.* group Ve-2	
ligenes type 2, group Va	*Pseudomonas-like* group 1, group Va-2	2	

are made from broth cultures. The Gram reaction and spore formation of an isolate suspected of being a pseudomonad should be carefully studied. Troublesome isolates presumptively considered to be *Pseudomonas* species may be gram-negative-staining *Bacillus* species or gram-positive coryneforms.

Screening Media

Kligler iron or triple sugar iron agar

Suspect *Pseudomonas* colonies on primary isolation media are transferred to Kligler iron agar (KIA) or triple sugar iron agar (TSIA) slants. Glucose-fermenting gramnegative rods produce an acid butt in these media. Glucose-nonfermenting gram-negative rods usually fail to produce visible growth in the butt, and there is no change in the color of the indicator in the butt after 24 h of incubation. Most glucosenonfermenting gramnegative rods produce an alkaline (red) slant. Strains of some species may produce no change in the color of the slant. Some *P. pseudomallei* strains and other *Pseudomonas* isolates may produce an acid (yellow) slant. *P. putrefaciens* causes the entire butt of KIA or TSIA to become black, without gas.

Isolates selected for further study should be streaked on an agar medium to verify the purity since a colony removed from primary isolation agar medium may not give rise to a pure culture.

Oxidative-fermentative basal medium

An isolate presumptively identified in KIA or TSIA as a glucosenonfermenting gramnegative rod should be confirmed as such by inoculating oxidative-fermentative (OF) basal medium with 1% glucose. The relatively high carbohydrate concentration of OF glucose medium permits the detection of oxidative acidity in the presence of ammonia and other bases derived from the. relatively small quantity of peptone in the medium. OF maltose, lactose, and xylose media should also be inoculated in the screening procedure. The media should be *pH 6.8* (green, not blue) at the time of inoculation. The colour of the indicator changes to green-yellow or yellow during an acid reaction and to blue during an alkaline reaction. No colour change indicates that the organism failed to grow.

Indophenol oxidase

To detect indophenol oxidase (cytochrome oxidase, cytochrome c oxidase) activity, growth can be pinked with a platinum loop and rubbed into a paper strip impregnated with oxidase reagent; the dark-blue color which appears within 30 s indicates indophenol oxidase activity. The test can also be performed by adding tetramethyl-p-phenylenediamine dihydrochloride or a-naphthol followed by dimethyl-p-phenylenediamine monohydrochloride reagents to colonies. The tetramethyl reagent tends to be more sensitive than the dimethyl reagent. Oxidase tests should be performed with 18- to 24-h-old growth by either method, using colonies present on noninhibitory medium, e.g., infusion agar. They should not be performed using colonies present on selective-differential agar media or Mueller-Hinton agar.

TABLE 6.2: MEDIA AND TESTS FOR ISOLATION AND IDENTIFICATION OF *PSEUDOMONAS* SPECIES

For isolation, presumptive identification, and screening Blood, infusion, or peptone agar

MacConkey, Leifson deoxycholate, or eosin-methylene blue agar

Kligler iron or triple sugar iron agar Motility

OF glucose, xvlose, lactose, and maltose media Indophenol oxidase

Arginine dihvdrolase (Moeller) Growth at 42°C

Odor, colonial morphology, pigmentation Somatic morphology (Gram stain) Flagella stain

Antibiogram

For further identification

Christensen urea agar Phenylalanine deaminase test agar Nitrate broth

Lysine and ornithine decarboxvlases DNase test medium OF fructose and mannitol media Esculin hydrolysis agar Nutrient gelatin

Tryptone broth

Acetate assimilation test medium

Arginine dihydrolase; lysine and ornithine decarboxylases

The activities of arginine dihydrolase and lysine and ornithine decarboxylases should be determined in Moeller decarboxylase base medium containing the indicated amino acid. After the media and the base medium control are inoculated, sterile mineral oil is aseptically overlaid. Enzymatic activity is indicated by an alkaline (dark purple) reaction when compared with the inoculated base medium control (light slate color).

Growth at 42°C

Growth at 42°C on an infusion agar slant is used in the differentiation of fluorescent *Pseudomonas* species which fail to produce pyocyanin.

Motility

Motility can be determined in a hangingdrop preparation which is prepared by placing a loopful of a broth culture in the center of a cover glass edged with petrolatum and inverting the cover glass on a concave glass slide. The preparations are examined with low- and high-dry objectives of the microscope.

Some glucose nonfermenters may fail to grow or spread in media used to detect motility of *Enterobacteriaceae*. Motility medium designed for glucose-nonfermenting gram-negative rods contains a critical quantity of agar and is also a convenient stock culture medium. The Gard plate, motility medium in a petri dish, is used to select a vigorously motile strain from a population of cells which appears nonmotile or sluggishly motile or which appears to contain few motile cells. The medium is stab inoculated in the center of the plate, and motile inoculum is later removed from the periphery of the giant spreading colony.

Flagellar morphology. Flagellar morphology is readily observed with a light microscope after staining, for example, with the tannic acid-fuchsin method described by Leifson. Although many *Pseudomonas* species can be identified and differentiated from other motile glucose-nonfermenting rods without determining their flagellar morphology, it should be determined for the identification of apyocyanogenic or melanogenic P. *aeruginosa* strains, motile glucose nonoxidizers (e.g., *P. alcaligenes, P. diminuta, P. testosterone, Alcaligenes faecalis, Bordetella bronchicanis)*, and enzymatically mutilated strains. Some glucosenonfermenting gram-negative rods cannot be identified without this information.

Most flagellated glucose-nonfermenting gram-negative rods can be divided into three groups based on attachment of the flagellum or flagella to the bacterial soma. Cells of strains of polar monotrichous species usually have one flagellum-per pole, and a few cells in the population may have two flagella at one pole. The number of flagella per pole on cells of strains of species with a tuft of polar flagella varies from zero to six or more. Cells with a flagellum or flagella at both poles often appear to be dividing. Cells of strains of peritrichous species have one or more flagella arranged on a cell. The flagellar shape and the site of attachment of the flagellum or flagella to the bacterial cell are usually uniform, constant, and characteristic of the species. Cells without flagella often occur in flagellated populations. Under certain conditions of growth, some strains produce morphovars with lateral flagella. The wavelength of lateral flagella is shorter than that of the terminal flagellum or flagella.

When *Pseudomonas* species are flagellated, the flagellum or flagella are attached at a pole. *Alcaligenes* species, *Bordetella bronchicanis, Agrobacterium* species, and *Achromobacter* species are peritrichous when flagellated, produce indophcnol oxidase, and arc asporogenous. *Acinetobacter* species are asporogenous nonmotile rods and do not produce indophenol oxidase, whereas *Moraxella* species are asporogenous nonmotile rods that produce indophcnol oxidase. *Flavobacterium* species are asporogenous nonmotile rods that produce indophenol oxidase, indole, and an intracellular yellow pigment. *Sphingobacterium* species do not produce indole but otherwise are similar to flavobacteria.

Additional Identification Methods

Approximately 15% of all gram-negative isolates from clinical specimens are glucose-nonfermenting gram-negative rods.

Pyocyanogenic *P. aeruginosa* isolates account for about 70% of these isolates. *Acinetobacter is* the second most frequently encountered glucose-nonfermenting gram-negative rod, followed by *P. maltophilia.* Other frequently encountered glucose nonfermenters include *P. fluorescens, P. putida, P. cepacia, P. stutzeri, Flavobacterium indologenes,* and *Alcaligenes* species.

Although some *Pseudomonas* species and other glucosenonfermenting gram-negative rods can be identified from information collected during the screening procedure described, it is necessary to detect additional characteristics) to identify other isolates.

Since *Pseudomonas* species do not produce indole, *Flavobacterium meningosepticum* and related species which produce indole can be distinguished. The synthesis of indole by glucose-fermenting gram-negative rods is usually detected by adding Kovacs reagent to a 24- or 48-h culture containing tryptone. Indole production by *Flavobacterium* species can be detected by adding Ehrlich reagent, after xylene extraction, to a 24-h broth culture. Urease activity can be detected quickly if the urea medium is warmed to incubator temperature before inoculation. The ability of pseudomonads to utilize an organic compound (e.g., acetate, DL-norleucine, pelargonate, testosterone, DL-β-hydroxybutyrate, p-hydroxybenzoate, etc.) as sole source of carbon and energy in a mineral base medium is employed in identification schemes.

Incubation Temperature

Although primary isolation media are usually incubated at 35 or 37°C, some *Pseudomonas* species in clinical specimens grow slowly at these temperatures and may be masked by other bacteria. Primary isolation media initially incubated at 35°C for 24 to 48 h should be reincubated at 30°C or room temperature (18 to 22°C) for an additional day or 2 to permit the growth of the temperature-sensitive bacteria. Media used to detect enzymatic activity of glucose-nonfermenting

TABLE 6.3: TYPE, NEOTYPE, AND LECTOTYPE STRAINS FOR IDENTIFICATION OF *PSEUDOMONAS* SPECIES

Species no.	*ATCC. no.*	*NCTCs status*	*Strain G + C*	*Mol%*
Pseudomonas cidovoran.s	15668	Type	67	
P.aeruginosa	10145	10332	Neotype	
P. alcaligenes	14909	10367	Neotype	66
P. cepacia	25416	10743	Type	67
P. diminuta	11568	8545	Type	67
P. fluorescens	13525	10038	Neotype	
P. gladioli	10248	Lectotype	68	
P. mallet	23344	Neotype	69	
P. inaltophilia	13637	10257	Type	67
P. mendocina	25411		Type	64
P. mesophilica	29983		Type	66
P. paucunobilis	29837	11030	Type	65
P. pertucinogena	190		Type	60
P. pickettii	27511		Type	64
P. pseudoalcaligenes	17440		Type	63
P. pseudomallei	1691		Neotype	
P. putida	12633		Neotype	
P. put refaciens	8071		Type	44
P. stutzeri	17588		Neotype	64
P. testosteroni	11996	10698	Type	62
P. vesicularis	11426		Type	66
Comamonas terrigena	8461	1937	Neotype	65

gram-negative rods should be incubated at 30°C unless otherwise stated or indicated. Since flagellar proteins are optimally synthesized at low temperatures, motility medium and cultures to be stained for flagella should be incubated at room temperature.

Each laboratory should establish in-house standards and data for identificationn of glucose-nonfermenting gram-negative rods. The identification procedure should be standardized to include a comparative study of the corresponding type, neotype, or hologype or a recognized reference strain of the species concerned. Some of the methods used in the study of pseudomonads have been described in greater detail.

A number of fractors determine the extent to which a laboratory should proceed with identification of glucose-nonfermenting gram-negative rods, including frequency of isolation, need for epidemiological surveillance, special interests of the infectious disease staff, and resources avilable to the diagnostic bacteriology laboratory, e.g. physical facilities, budget, knowledgeable personnel, and time. Some diagnostic laboratories may find it expedient to use one or more of teh commercial micromethod identification kits designed for glucose nonfermenters; other may find it expedient to send suspected *Pseudomonas* species and other glucose-nonfermenting gram-negative rods to a reference laboratory for further identification.

PSEUDOMONAS SPECIES

The minimal characters for identification of most strains of *Pseudomonas* species are recorded in Table 4. The species are obligate aerobes and have a respiratory type of metabolism with oxygen as the terminal electron acceptor. Some species can use nitrate as an alternative electron acceptor and grow under anaerobic conditions in the presence of nitrate. Energy is not obtained by fermentative or photosynthetic metabolism. Some species do not produce acid from glucose. Indole and acetylmethylcarbinol are not produced, and the methyl red test is negative. *Pseudomonas* species produce various fluorescent and nonfluorescent pigments. Most species grow in a mineral base medium, without organic growth factors, containing ammonium ion as sole source of nitrogen and glucose as sole source of carbon and energy. A few species require vitamins or amino acids as growth factors. Most pseudomonads contain *57* to *70* mol% guanine plus cytosine (G+C) in their DNA. The characters of *Pseudomonas* species and the minimal characters for identification of each are based on a study of 2,102 strains.

FLUORESCENT GROUP

Most strains of the three species of the fluorescent group produce water-soluble pigments (pyoverdins) which fluoresce when exposed to UV light of short wavelength (ca. 254 nm). Fluorescent pigments may be yellow-green, yellow-brown, or without color. They are not soluble in chloroform. Production of pyoverdins is influenced by nutritional factors, and media which support growth may not promote their synthesis. Most strains produce indophenol oxidase and arginine dihydrolase. Those strains which lack these features may be more closely related to phytopathogenic fluorescent pseudomonads such as *P. syringae*. The majority of strains grow in mineral base medium containing ammonium ion as sole source of nitrogen and glucose as sole source of carbon and energy.

PSEUDOMONAS AERUGINOSA

Identification

Most *P. aeruginosa* strains are identified on the basis of the characteristic grapelike odor of aminoacetophenone, colonial morphology,. and production of pyocyanin, a blue, water-soluble, nonfluorescent, phenazine pigment that is soluble in chloroform. *P. aeruginosa is* the only *Pseudomonas* species or gramnegative rod known to excrete pyocyanin. Strains which produce pyocyanin and the yellowish pyoverdins often impart a greenish color to culture media.

P. aeruginosa strains may synthesize various combinations of pyocyanin, pyoverdins, pyorubin (red, water-soluble pigment), and pyomelanin (brown to black, water-soluble pigment). Pyocyanin may

TABLE 6.4: MINIMAL CHARACTERS FOR IDENTIFICATION OF STRAINS OF *PSEUDOMONAS* SPECIES

Character	*Sign*
Gram-negative, straight or slightly curved rod	+
AsporogenousPolar monotrichous or polar tuft of flagella"	+
Motility°	+
OF glucose medium open, acid	+ or
OF glucose medium sealed, acidIndophenol oxidase	+ or –
Catalase	+
Photosynthetic pigment	–

be masked by other pigments. Strains which produce pyorubin or pyomelanin, as well as highly mucoid strains, may fail to oxidize carbohydrates, as do most strains of *P. aeruginosa*. Highly mucoid strains may also lack motility.

A number of colony types (smooth, coliform-type, rough, mucoid, gelatinous, dwarf) may be observed. Colonial variants arising by dissociation of a single strain give the false impression that different bacterial species are present. The variant forms may demonstrate altered exoenzyme activity, bacteriophage pattern, serovar, antibiograms, and antigenicity.

TABLE 6.5: MINIMAL CHARACTERS FOR IDENTIFICATION OF APYOCYANOGENIC *P. AERUGINOSA* STRAINS

Character	Sign	% Positive
Polar monotrichous, fewer than three flagella per pole	+	*97*
Motility	+	*97*
OF glucose medium open, acid	+	*98*
OF lactose, sucrose media, acid	0	
Indophenol oxidase	+	100
Pyoverdin	+ or -	*71*
L-Arginine dihydrolase	+	*99*
Growth at 42°C	+	100

Apyocyanogenic strains may produce aberrant biochemical character patterns, and their identification may pose a problem. The following universal characters of pyocyanogenic strains are unreliable for identification of apyocyanogenic strains: grapelike odor; nitratase, nitritase, gelatinase, caseinase, and lipase activity; acid production from mannitol in OF medium; utilization of acetamide; tolerance to cetrimide; and susceptibility to carbenicillin. Apyocyanogenic strains are more often susceptible to tetracycline, streptomycin, and kanamycin than are pyocyaninproducing strains. Generally, nonpigmented strains are not typable by antisera usually used to determine the serovar of *P. aeruginosa* strains, and they show little or no aeruginocin (pyocin) activity. Apyocyanogenic strains can be recognized as variants of *P. aeruginosa* by several uniform characters, namely, the presence of polar monotrichous flagella, growth at 42°C, and failure to produce acid from lactose and sucrose. Strains designated as unidentified fluorescent *Pseudomonas* species are similar to apyocyanogenic *P. aeruginosa* strains in that they grow at 42°C, but they possess a polar tuft of flagella.

Chemical constituents of bacterial cells, such as cellular fatty acids, respiratory quinone systems, and biosynthetic enzymes, are potentially useful tools for identification of *P. aeruginosa* and other *Pseudomonas* species. Serological, bacteriophage, aeruginocin typing, and enzyme profile systems have been used as epidemiological markers.

Sources of Isolation and Relationship to Disease

P. aeruginosa is a physiologically versatile organism, widely distributed in soil, water, sewage, the mammalian gut, and plants. It may be pathogenic for humans, as well as for certain animals, insects, and plants. *P. aeruginosa is* susceptible to drying but will survive for many months in water at ambient temperature. Simple nutritional requirements and the ability to metabolize a variety of organic substances may enable *P. aeruginosa* to survive and multiply in fluids and moist environments found in hospital wards.

Infection with *P. aeruginosa is* generally restricted to hospitalized patients. Predisposing factors for nosocomial infections include metabolic, hematologic, and malignant diseases. Hospital-acquired infections occur in patients who have had prior instrumentation or manipulative procedures such as urethral catheterizations, tracheostomies, lumbar punctures, and intravenous infusions of medications and fluids. Patients become susceptible to *P. aeruginosa* infections after prolonged treatment with immunosuppressive agents, corticosteroids, antimetabolites, antibiotics, and radiation. Hospitalized patients may acquire the bacteria from common environmental sources by contact with human or inanimate vectors. Sources in the hospital environment shown to be transitory vectors include infusion fluids, ophthalmic solutions, dilute phenolic solutions, benzalkonium chloride solution, hexachlorophene soap, hand creams, body lotions, bedside water decanters, ornamental plants, flower vase water, forceps, shaving brushes, oral thermometers, suction apparatus, respiratory ventilators, nebulizing humidifiers, incubators, whirlpools, syringes, ward utensils, dish mops, sponges, floors, baths, sinks, and water taps.

Alternatively, patients may develop an endogenous infection from bacteria present on their skin or in their respiratory or intestinal tract. The proportion of the general population carrying *P. aeruginosa* in the intestinal tract is low, but the administration and selective influence of antibiotic therapy among hospitalized patients have increased the carriage rate among these subjects. The most important human sources for *P. aeruginosa* are infected wounds, urine, and lesions producing an exudate. These moist sources contaminate dressings, bedsheets, patient garments, and equipment, from which the organism may be transferred to other patients by the hands of hospital personnel or visitors. Person-toperson transfer appears to be the most important mode of transmission. *P. aeruginosa is* a cause of severe epidemic diarrhea of infants, ocular infections, burn infections, cystic fibrosis, hot tuband whirlpool-associated folliculitis, osteomyelitis, and malignant external otitis. Susceptibility tests of *P. aeruginosa* by broth dilution methods may yield low MICs with the aminoglycosides because of the low concentration of magnesium and calcium ions in Mueller-Hinton and many other broth media. Agar diffusion procedures yield more reliable results because of the optimal concentration of divalent cations contributed by most agar media.

TABLE 6.6: MINIMAL CHARACTERS FOR IDENTIFICATION OF *P. FLUORESCENS* AND *P. PUTIDA* STRAINS

	P. fluorescens		*P. putida*	
CharacterSign	***%***	***Positive***	***Sign %***	***Positive***
Polar tuft of three or	+	100	+	100
more flagella perpoleMotility	+	**100**	+	100
OF glucose medium	+	100	+	100
open, acidOF lactose medium,	– or +	12	– or +	14
acidIndophenol oxidase	+	100	+	100
Pyocyanin	0	0		
Pyoverdin	+	95	+ or –	84
L-Arginine dihydrolase	+	99	+	99
Gelatin hydrolysis	100	0		
Growth at 42°C	0	0		

P. aeruginosa produces a variety of toxins and enzymes such as surface slime, hemolysin, fibrinolysin, lipase, esterase, lecithinase, elastase, coagulase, DNase, phospholipase, endotoxin, enterotoxin, and exotoxin, some of which may contribute to its pathogenesis. The role of virulence factors in the pathogenesis of *P. aeruginosa* has been demonstrated in animal models. A number of literature reviews summarize the relationship of *P. aeruginosa* and other pseudomonads to human diseases.

Pseudomonas Fluorescens and *Pseudomonas Putida*

Identification

Strains of *P. fluorescens* and *P. putida* have a polar tuft of flagella. The number of flagella per cell varies from zero to six or more. Most strains contain some cells with three or more flagella per pole. The number of flagella per pole is a valuable criterion for distinguishing *P. aeruginosa from P. fluorescens* and *P. putida. P. aeruginosa* strains are polar monotrichous and very rarely contain cells with three flagella per pole. *P. fluorescens* and *P. putida* do not produce pyocyanin, and some strains fail to produce fluorescent pigments. They do not grow at 42°C, and many strains fail to grow at 35°C.

These two species historically have been distinguished by proteolytic activity. *P. fluorescens,* but not *P. putida,* produces an extracellular protease, a metalloenzyme, which hydrolyzes gelatin and casein; there is some question, however, concerning the reliability of distinguishing these strains on this basis alone. The variable production of lipase (formation of calcium oleate from Tween 80) and hydrolysis of phosphatidylcholine (egg yolk reaction) by *P. fluorescens* help distinguish it from *P. putida. P. putida* does not reduce nitrate to nitrogen gas. *P. fluorescens* and *P. putida* are distinguished from apyocyanogenic strains of *P. aeruginosa* by flagellar morphology, growth at 42°C, and acid production in OF medium from such disaccharides as lactose, maltose, and sucrose. Whereas *P. fluorescens* and *P. putida* are usually susceptible to kanamycin and resistant to carbenicillin, the reverse is true for most *P. aeruginosa* strains.

P. fluorescens strains can be subdivided into seven biovars (A through G), and *P. putida* strains can be subdivided into two biovars on the basis of denitrification, pigment production, levan synthesis from sucrose, and utilization of certain organic compounds. Three phenazine-producing biovars of *P. fluorescens (P.* chlororaphis, *P.* aureofaciens, *P.* lemonnieri) phenotypically resemble *P. fluorescens* but are recognized as distinct species by some workers.

Sources of Isolation and Relationship to Disease

P. fluorescens and *P. putida* are common inhabitants of soil, water, and plants. They have been recovered from the skin and ears of divers; cosmetics; aviation turbine fuel; hard and soft lens wetting, cleaning, and saline solutions; and contaminated foodstuffs including beef, pork, and milk. They have been frequently isolated from the hospital environment, i.e., water sources, sink drains, floors, benzalkonium chloridesolution, fluid from an ultrasonic nebulizer, humidifiers, intravenous catheter, and contaminated blood and blood products stored in a refrigerator.

The species are part of the normal oropharyngeal flora, and most strains are isolated in the clinical laboratory from respiratory tract specimens. A number of these isolates are from cystic fibrosis patients. These organisms are usually environmental contaminants and are rarely opportunistic pathogens for humans. *P. fluorescens* can release endotoxins in contaminated blood and blood products. The phospholipid moiety of the endotoxin produces irreversible shock. These bacteria have been associated with empyema, urinary tract infection, pelviperitonitis, infection in a chronic bronchitic, wound infections, septicemia, and septic arthritis.

STUTZERI GROUP

The two species in the stutzeri group are salt tolerant and nonhalophilic but require sodium cation for growth. They grow under strictly anaerobic conditions in nitrate-containing media, as do *P. aeruginosa, P. pseudomallei,* and other bacterial species which reduce nitrate to nitrogen gas. The strains are motile with polar monotrichous flagella. Lateral flagella are produced by some cells in some strains. Indophenol oxidase and oxidative acidity from glucose are produced. These two species grow with ammonium ion as sole source of nitrogen and acetate as sole source of carbon and energy. They are susceptible to polymyxins.

PSEUDOMONAS STUTZERI

Identification

P. stutzeri has the characters of the genus *Pseudomonas* as recorded in Table 4. The characteristics of 168 strains of *P. stutzeri* are recorded in. Minimal characters for identification of most *P. stutzeri* strains are recorded in. A neotype strain has been designated and described.

Most freshly isolated strains produce dry, wrinkled, leathery, adherent colonies as well as smooth colonies and various intermediate types. Colonies may become buff to light-brown in color due to high concentrations of cytochrome c in the cells. Smooth colonies predominate after repeated subculture. Rare strains produce only smooth, spreading colonies, similar to strains of *Proteus* species, or colonies which pit blood agar medium.

Ability to produce nitrogen gas may be lost on repeated subculture. Strains which no longer denitrify or produce wrinkled colonies will regain these features after serial passage in nitratecontaining media. Arginine dihydrolase is produced only by rare, biovar Vb-3 strains.

Maltose and, usually, starch are utilized. These and other reactions distinguish *P. stutzeri* from *P. mendocina*. *P. stutzeri* may be confused with *P. pseudomallei*, since both produce rough, wrinkled, corrugated colonies. Flagellar morphology; susceptibility to polymixin; starch, arginine, and lactose reactions; and other characters distinguish *P. stutzeri* from *P. pseudomallei*.

P. stutzeri demonstrates marked nutritional versatility in that only 17 organic compounds serve as substrates for all strains, whereas 72 compounds can be used by some strains. The wide variation in the DNA base composition, 60.7 to 66.3 mol% G+C, of *P. stutzeri* may be related to the high degree of nutritional heterogeneity. Since the phenotypic characters of *P. stutzeri* are heterogeneous, it may include several biologically distinct species.

Sources of Isolation and Relationship to Disease

P. stutzeri is ubiquitous in soil and water and has been found in manure, humus, straw, sewage, stagnant water, distilled water, deionized water, sink drains, tap water aerator, baby formula, aerosolization equipment, hemodialysis machine, inanimate surfaces, aqueous green soap, hard lens wetting solution, eye cosmetics,, and various clinical specimens. On a few occasions P. *stutzeri* has been directly associated with an infectious process, e.g., endocarditis, septicemia, septic arthritis, postoperative and posttraumatic infections of the extremities, otitis media, conjunctivitis, and urinary tract infections.

PSEUDOMONAS MENDOCINA

Identification

P. mendocina has the characters of the genus *Pseudomonas* as recorded in. The characteristics of six strains of *P. mendocina* are recorded. Minimal characters for identification of most *P. mendocina* strains are recorded in. *P. mendocina* ATCC 25411 (CH_2O) should be recognized as the type.

Colonies of *P. mendocina* are flat, smooth, and butyrous and produce a brown-yellow, intracellular, carotenoid pigment. Wrinkled colonies are not produced. Water-insoluble, alcoholsoluble carotenoid and carotenoidlike pigments, such as those found in *P. mendocina,* characteristically remain associated with the cells. Arginine dihydrolase is produced. Starch and maltose are not utilized. The base variation of the DNA in *P. mendocina* shows little variation, with 62.8 to 64.3 mol% G+C content.

TABLE 6.7: MINIMAL CHARACTERS FOR IDENTIFICATION OF *P. STUTZERI* AND *P. MENDOCINA* STRAINS"

Character	*P. stutzeri*		*P. mendocina*	
	Sign %	Positive	Sign%	Positive
Polar monotrichous, fewer than threeflagella per poleMotility	+	100	+	100
	+	100	+	100
OF glucose medium open, acid	+	100	+	100
OF lactose medium, acid	–	0	–	0
OF maltose medium, acid	+	98	–	0
Indophenoloxidase	+	100	+	100
Nitrate to gas	+	100	+	100
t.-Arginine dihydrolase	–	5	+	100

P. mendocina may be confused with denitrifying biovars of *P. fluorescens*. Polar monotrichous morphology, growth at 42°C, and other characters distinguish it from nitrogen gas-producing strains of P. fluorescens.

Sources of Isolation and Relationship to Disease

P. mendocina has been isolated from soil, water,. and urine. It has not been associated with infections in humans.

ALCALIGENES GROUP (PSEUDOMONAS ALCALIGENES AND PSEUDOMONAS PSEUDOALCALIGENES)

The two species in the alcaligenes group have the characters of the genus *Pseudomonas*. The cells of most strains have no somatic curvature, and some strains contain a few cells which are distinctly curved. They are polar monotrichous when flagellated, and the wavelength of the flagellum is approximately 1.6 gm. The flagellar morphology of the species in the alcaligenes group is similar to that of *P. aeruginosa*.

Most strains grow in mineral base medium containing ammonium ion as sole source of nitrogen and acetate as sole source of carbon and energy. The neotype strain of *P. alcaligenes* does not utilize glucose, and alkali accumulates in unsealed OF glucose medium and OF basal medium. Indophenol oxidase is produced. Pyocyanin and pyoverdins are not produced.

PSEUDOMONAS ALCALIGENES AND PSEUDOMONAS PSEUDOALCALIGENES

Identification

The characteristics of 51 strains of *P. alcaligenes* and 62 strains of *P. pseudoalcaligenes* are recorded. Minimal characters for identification of most *P. alcaligenes* and *P. pseudoalcaligenes* strains are recorded in. The neotype strain of *P. alcaligene* and the type strain of *P. pseudoalcaligenes* have been described.

Strains assigned to the alcaligenes group can be accommodated in two named species, *P. alcaligenes (P. alcaligenes* biovar A) and *P. pseudoalcaligenes (P. alcaligenes* biovar *B). P. pseudoalcaligenes* produces a weak acid reaction from fructose in OF basal medium and is nutritionally more versatile than *P. alcaligenes.* Fructose is the only substrate used by all strains of *P. pseudoalcaligenes* and none of the strains of *P. alcaligenes.* An additional five substrates not used by any of the latter are used by most of the more versatile *P. pseudoalcaligenes* strains. Phenotypic characters do not easily distinguish the two species. Hybridization experiments indicate a high degree of homology within each of the two species and a clear separation between them.

Sources of Isolation and Relationship to Disease

P. alcaligenes and *P. pseudoalcaligenes* have been isolated from pond, river, swimming pool, and turtle aquarium waters; cosmetics; cell culture; raw milk; frozen fish; liver of piglet; rabbit blood culture; feces of rabbits, frogs, and humans; and numerous clinical sources. They are occasionally opportunistic agents. *P. alcaligenes* has been associated with endocarditis, neonatal septicemia, empyema, and eye infections. *P. pseudoalcaligenes* has been identified as the etiological agent of pneumonitis, a postoperative knee infection, septicemia, and meningitis.

Pseudomonas sP. group 1, which contains some strains previously designated *Pseudomonas denitrificans, is* distinguished from *P. alcaligenes* by the reduction of nitrate to nitrogen gas. The species name P. *denitrificans* Bergey et al. *1923* has been rejected by the Judicial Commission of the International Committee on Systematic Bacteriology, since strains designated *P. denitrificans* appear to belong to species in several genera. An isolate referred to as *P. denitrificans* caused bacteremia and meningitis.

PSEUDOMALLEI GROUP

The principal feature shared by the species in the pseudomallei group is nutritional versatility in the type and number of organic compounds utilized as sole sources of carbon and energy, which include carbohydrates, mono- and dicarboxylic acids, monoand polyalcohols, aromatic compounds, amino acids, and amines. Growth occurs in mineral base medium containing ammonium ion as sole source of nitrogen and glucose as sole source of carbon and energy. A few strains fail to produce indophenol oxidase, and others produce a slow and very weak indophenol oxidase reaction. The majority of strains are not susceptible to antibiotics of the polymyxin class.

TABLE 6.8: MINIMAL CHARACTERS FOR IDENTIFICATION OF *P. ALCALIGENES* AND *P. PSEUDOALCALIGENES*

Character	***P. alcaligenes caligenes***		***P. pseudoal***	
	Sign%	**Positive**	**Sign%**	**Positive**
Polar monotrichous, fewer than three flagella perpole, and normal wavelength	+	100	+	100
Motility	+	100	+	100
OF glucose medium open,	-	0	· or +	12
acidOF fructose medium, acid	-	0	+	100
Indophenol oxidase	+	100	+	100
Hydrogen sulfide, black butt in KIA	0	-	0	

P. pickettii, the animal pathogens *P. mallei* and *P. pseudomallei,* and the plant pathogens *P. cepacia, P. gladioli (P. marginata, P. alliicola),* and P. *caryophylli* are enzymologically, phenotypically, and genotypically related.

PSEUDOMONAS PSEUDOMALLEI

Identification

P. pseudomallei has the characters of the genus *Pseudomonas* as recorded in. The characteristics of seven strains of *P. pseudomallei* are recorded in. Characters useful for the identification of *P. pseudomallei* strains are recorded in. The neotype strain was isolated in 1921

Colonies vary from mucoid and smooth to rough and wrinkled in texture and from bright orange to cream color. Colonial dissociation occurs with the same frequency as in P. *aeruginosa. P. pseudomallei, P. mallei, P. stutzeri,* and *P. aeruginosa,* as well as other pseudomonads, may produce wrinkled, rough, corrugated colonial variants. Initial growth *of P. pseudomallei is* accompanied by an odor of putrefaction supplemented by a distinctive, aromatic, pungent odor. Strains are motile with a tuft of three or more flagella per pole. Abundant growth occurs under anaerobic conditions in nitrate-containing media. Gelatin is hydrolyzed, lactose is utilized, and growth

TABLE 6.9: CHARACTERS FOR IDENTIFICATION OF *P. PSEUDOMALLEI* AND *P. MALLEI* STRAINS

	P. pseudo-mallei		***P. mallet***	
Character	**Sign**	**% Positive**	**Sign**	**% Positive**
Polar tuft of three or	+	100		0
more flagella per poleMotility	+	100		0
OF glucose medium	+	100	+	**100**
open, acidOF maltose medium, acid		+	100	(+)
100				
Citrate, Simmons	+	96		0
Nitrate to gas	+	100		0
L-Lysine decarboxylase	-	0	-	0
L-Arginine dihydrolase	+	100	+ or (+)	100
Growth at 42°C	100		0	

Sources of Isolation and Relationship to Disease

P. pseudomallei causes melioidosis, an endemic glanders-like disease of humans and animals, in Southeast Asia and northern Australia. Human melioidosis rarely occurs in the Western Hemisphere. It usually occurs in the United States among those who have returned from countries where it is endemic. The free-living soil organism usually infects the host through a skin opening or,by inhalation after the host has had contact with soil or water. A soiled pelvic wound received by a man in a farming accident in Oklahoma became infected with an organism identified by cultural and biochemical characteristics as *P. pseudomallei.* It is rarely transmitted from person to person. Prostatic melioidosis can be the source of human-to-human sexual transmission of P. *pseudomallei*. The organism may remain dormant, and asymptomatic infection can persist in humans for many years before the appearance of clinical disease. The incidence of latent infections in humans in endemic areas appears high. Pulmonary melioidosis is a relatively benign disease with good prognosis. Untreated septicemic melioidosis progresses rapidly and has a high mortality rate. Patients in Australia developed nosocomial infection after urethral catheterization; the source of the organism was probably soil on the hospital grounds. Laboratory-acquired infections have been reported. Because the host spectrum is broad and the clinical manifestations of melioidosis in humans are varied, diagnosis is dependent on isolation and identification of the agent of infection as P. pseudomallei.

PSEUDOMONAS MALLEI

Identification

P. mallei has the characters for the genus *Pseudomonas* as recorded in. It is the only nonmotile species in the genus *Pseudomonas*. The characteristics of six strains of *P. mallei* are recorded in. The neotype strain has been described.

Colonies are smooth and range from white to cream color. *P. mallei* grows slowly as compared with *P. aeruginosa* or *P. pseudomallei,* and the cell crop yield is lower. It fails to produce nitrogen gas, hydrolyze gelatin, or grow at 42°C. Although *P. mallei* vigorously reacts with *P. pseudomallei* antiserum, it is distinguished from *P. pseudomallei* by the characters recorded in.

Sources of Isolation and Relationship to Disease

P. mallei is the cause of glanders (farcy), a natural disease primarily of equines. Occasionally, it is transmitted from equine hosts to humans by direct contact through abraded skin and inhalation. It can be transmitted from person to person. Glanders has been eradicated in the United States and Canada. *P. mallei* can be isolated from blood, sputum, or pus from patients with glanders. It is thought to be the only highly adapted parasite of animals in the genus Pseudomonas.

PSEUDOMONAS CEPACIA

Identification

P. cepacia has the characters of the genus *Pseudomonas*. The characteristics of 201 strains of *P. cepacia* are recorded in. Minimal characters for identification of most *P. cepacia* strains are recorded in. The type strain has been designated.

Some *P. cepacia* strains produce a conspicuous sulfur-yellow or green, water-soluble, chloroform-soluble, nonfluorescent phenazine pigment. On chemically defined media, the pigments may show a great variety of colors including brown, red, and purple, depending on the carbon sources used for growth. A strain may produce one or more pigments. These pigments occur in the bacterial growth and often diffuse into the agar medium. A diffusible fluorescent pigment may be produced which is violet in UV light and is distinguished from the bluish-green fluorescence of *P. fluorescens*. Many strains of clinical origin are nonpigmented.

TABLE 6.10: MINIMAL CHARACTERS FOR IDENTIFICATION *OF P. CEPACIA* STRAINS

Character	*Sign*	*% Positive*
Polar tuft of three or more flagellaper pole	+	99
Motility	+	99
OF glucose medium open, acid	+	100
OF lactose medium, acid	+	98
OF maltose medium, acid	+	97
OF mannitol medium, acid	+	100
Nitrate to gas	0	
L-Lysine decarboxylase	+	90
L-Arginine dihydrolase	0	

Colonies on blood agar are opaque, glistening, occasionally mottled, convex in elevation, and butyrous in consistency. A sweet odor characteristically associated with *P. aeruginosa* is produced by some strains. Isolates should be subcultured frequently since they tend to become nonviable after 2 to 3 days on blood agar medium. Optimal growth occurs at 30 C, and some strains grow poorly at 35°C.

P. cepacia is motile with a tuft of three to eight polar flagella. Some cells in a population may have no flagella, and some may have only one flagellum per pole. Cells with lateral flagella occur. A few strains are atrichous. o-Nitrophenyl-13-D-galactopyranoside (ONPG), lysine decarboxylase, esculin hydrolysis, proteolytic activity, acidification of carbohydrates, and other reactions distinguish *P. cepacia* from other pseudomonads. A classification into at least eight biovars based on ONPG, nitratase, and esculin hydrolysis permits epidemiological marking of the strains. Bacteriocin patterns and plasmid profiles distinguish strains of clinical and plant origin. A clinical serotyping scheme is based on 12 somatic and flagellar antigens.

TABLE 6.11: MINIMAL CHARACTERS FOR IDENTIFICATION OF *P. PICKETTII* STRAINS

Character	*Biovar 1*		*Biovar 2*	
	Sign	**% Positive**	**Sign**	**% Positive**
Polar monotrichous, fewer than three flagella per pole	+	96	+	100
Motility	+	96	+	100
OF glucose medium open, acid	+ or (+)	100	+ or (+)	100
OF lactose medium, acid	+ or (+)	100	–	0
OF maltose medium, acid	+ or (+)	100	–	0
OF mannitol medium, acid	–	0	–	0
Indophenol oxidase	+	100	+	100
Nitrate to gas	+, (+) or –	88	+ or (+)	100
L-Arginine dihydrolase	–	0	–	0

It is likely that some clinical isolates identified as *P. cepacia* more closely resemble *P. gladioli*. The latter species gives negative reactions for indophenol oxidase, oxidation of sucrose and maltose, and lysine decarboxylase tests, and a positive reaction for urease test.

Isolation and Relationship to Disease

P. cepacia was first described as a phytopathogen that causes sour skin, an onion bulb rot. It has a wide geographic distribution and has been isolated from raw and pasteurized milk; hard lens and soft lens wetting and saline solutions; nasal sprays; rotting tree trunk; forest soil, muck soil, and soil enrichments; and natural, river, and tap waters. In hospital, *P. cepacia* has been isolated from distilled water, flower vases, disinfectants, detergent solutions, saline solution, aerosol polymyxin, spray atomizer, anesthetics, thermometers, ultrasonic nebulizers, incubators, respirators, humidifiers, water bath, sink drains, pressure transducers, intravenous cannula, hemodialysis coils, hand dip, urinary catheter kits, baby lotion, intravascular inserts, and intravenous fluids.

P. cepacia has been isolated from a variety of clinical materials. It is an opportunistic pathogen, and most infections are of nosocomial origin. *P. cepacia* has been associated wth septicemia, meningitis, endocarditis, urinary tract infections, pneumonia, postoperative wound infections, abscesses, septic arthritis, peritonitis, conjunctivitis, chronic granulomatous disease, and cystic fibrosis. *P. cepacia* demonstrates a multiple antibiotic resistance pattern, especially to the aminoglycosides usually active against *P. aeruginosa.*

Intraperitoneal injection of *P. cepacia* cells in mice resulted in 50% o lethal dose values of *3.8* x 10^8 cells. Endotoxin can be detected by the *Limulus* amoebocyte lysate assay. Animal inoculation is not reliable in the differentiation of *P. cepacia* from *P. pseudoinallei,* since some isolates of *P. cepacia* produce pathological changes in animals similar to those produced by P. *pseudomallei.*

PSEUDOMONAS PICKETTII

Identification

Cells of strains of P. *pickettii* are polar monotrichous. Occasional cells with two polar flagella are observed. The development of growth on blood agar medium characteristically is slow. Growth is not pigmented, and soluble pigments are not produced. Growth on blood agar medium tends to become nonviable after *3* or *4* days.

The development of acid reactions from carbohydrates in OF basal medium and the reduction of nitrate to nitrogen gas characteristically are slow and may require *48* h of incubation to detect. The optimum temperature of incubation for the detection of nitrogen gas is *30°C,* not *35*°C. Strains which reduce nitrate to nitrogen gas grow anaerobically in the presence of nitrate ion and an oxidizable substrate.

Urease and indophenol oxidase are produced. Arginine dihydrolase and ornithine and lysine decarboxylases are not produced. Strains grow in mineral base medium containing ammonium ion as sole source of nitrogen and glucose as sole source of carbon and energy.

The morphology and the delayed denitrification and delayed carbohydrate oxidation reactions may have contributed to the misidentification of these strains as *P. alcaligenes* or *P. pseudoalcaligenes.*

P. pickettii is a distinct but heterogeneous species and comprises up to seven biovars based on different oxidation patterns and other phenotypic features, i.e., acid production from lactose, maltose, and mannitol and gelatinase production *(33, 56). P. pickettii* biovar 1 (CDC group Va-1) produces acid from lactose and maltose, whereas biovar 2 (CDC group Va-2) does not.

Isolation and Relationship to Disease

P. pickettii has been isolated from patient sources and from environmental sources including water from an artificial kidney machine, autoclave-cooling water, intravenous fluids, umbilical catheter tip, intravenous catheter, respiratory therapy saline solution, dialysis fluid, nebulizers, cosmetics, paper mill effluent, and teething rings. *P. pickettii* has been associated with septicemia, a case of acute, nonfatal meningitis, and infections of the urinary and respiratory tracts.

ACIDOVORANS GROUP

The species in the acidovorans group have the characters of the genus *Pseudomonas* are recorded, and they also. share other attributes. Although most strains are populations of cells with no somatic curvature, an occasional cell of some strains is distinctly curved. Cells are motile with a polar tuft of flagella. The number of flagella at one pole varies from one to six. Strains usually contain some cells with tufts which possess three or more flagella. The tuft of flagella is usually at one pole of the cell; cells with tufts at both poles appear to be dividing. Polar flagella have a mean wavelength of 3.1 Km and an amplitude of 1.08 gm. This distinctive flagellar shape has been described and illustrated in photomicrographs. Some strains produce cells with lateral flagella.

Growth is not pigmented, but colonies of some strains are surrounded by a tan or brown zone of discolored agar medium. Most strains grow in mineral base medium containing ammonium ion as sole source of nitrogen and acetate as sole source of carbon and energy.

Alkali accumulates at the surface of OF glucose medium. Acid is not produced from glucose in OF basal medium, and most carbohydrates are not oxidized. Indophenol oxidase is produced. Although indole is not produced, some strains produce anthranilic acid and kynurenine in tryptone broth. These and other tryptophan derivatives cause Kovacs reagent to become orange.

Pseudomonads in the acidovorans group, e.g., *P. acidovorans* and *P. testosteroni,* have the characteristics which Guenther attributed to *Vibrio terrigenus*. During the past *90* years, numerous designations, including *Pseudomonas terrigena,* have been used for this species. The specific epithet in the name *Vibrio terrigenus* has priority. The curved soma, particularly the multiple curves of rough and filamentous forms, flagellar morphology, and physiology suggest a relationship of this taxon to certain spirilla. Electron microscopy shows a layer of regularly arranged subunits on the outer surface of the cell wall of *P. acidovorans*. The subunits are similar to those of *Spirillum* species, but they do not occur in other pseudomonads. Hugh proposed the new combination *Comamonas terrigena* for this taxon. *ATCC* 8461 is the neotype strain.

PSEUDOMONAS ACIDOVORANS AND PSEUDOMONAS TESTOSTERONI

The characteristics of 95 strains of *P. acidovorans* and 18 strains of *P. testosteroni* are recorded. Minimal characters for identification of most *P. acidovorans* and *P. testosteroni* strains are recorded. The type strains of *P. acidovorans* and *P. testosteroni* have been described.

TABLE 6.12: CHARACTERS USED IN IDENTIFICATION OF ACIDOVORANS GROUP *PSEUDOMONAS* SPECIES

Test, substrate, or morphology	*P. acidovorans (95 strains)*		*P. testosteroni (18 strains)*	
	Sign%	*Positive*	*Sign%*	*Positive*
Acid:Glucose, 1% (OFBM)	—	0	—	0
Fructose	+	100	—	0
Galactose	—	0	—	0
Mannose	—	0	—	0
Rhamnose	—	0	—	0
Xylose	—	0	—	0
Lactose	—	0	—	0
Sucrose	—	0	—	0
Maltose	—	0	—	0
Mannitol	+	95	—	0
Lactose, 10% (PAB)	—	0	—	0
ONPG	—	0	—	0
Pyoverdin	—	0	—	0
Hydrogen sulfide (KIA)	—	0	—	0
Nitrate reduction	+	95	+	94
Gas from nitrate	—	0	—	0
Indophenol oxidase	+	100	+	100
Arginine dihydrolase	—	0	—	0
Lysine decarboxylase	—	0	—	0
Ornithine decarboxylase	—	0	—	0
Phenylalanine deaminase	—	3	—	6

Hydrolysis:				
Urea	—	0	—	0
Esculin	—	0	—	0
Tween 80	- or +	20	- or +	17
Starch	—	0	—	0
DNA	—	1	—	0
Lecithin	—	0	—	0
Gelatin	—	2	—	0
Acetamideb	+	100	—	0
Hemolysis	—	0	—	0
Growth on:SS	+or-	66	—	6
MacConkey	+	99	+ or -	89
6.5% NaCl	—	0	—	0
Cetrimide (PA)	- or +	12	—	6
MBM + acetate	+	100	+ or -	67
Growth at 42°C	-or +	10	+ or -	50
Polymyxin susceptible	+ or -	61	+	94
Motile	+	100	+	100
No. of flagella		> 1		>1

TABLE 6.13: MINIMAL CHARACTERS FOR IDENTIFICATION OF *P. ACIDOVORANS* AND *P. TESTOSTERONI* STRAINS

Character	*P. acidovorans*		*P. testosterone*	
	Sign %	Positive	Sign %	Positive
Polar tuft of three or more flagella per pole, and long wavelength	+	100	+	100
Motility	+	100	+	100
OF glucose medium open,acid	-	0	-	0
OF fructose medium, acid	+	100	-	0
OF mannitol medium, acid	+	95	-	0
Indophenol oxidase	+	100	+	100
Hydrogen sulfide, black butt in KIA	-	0	-	0

Pseudomonads in the acidovorans group are divided into two specie's on the basis of utilization of fructose and additional phenotypic and genotypic differences. Non-fructose-utilizing *P. testosteroni* strains arc distinguished from fructose-utilizing *P. acidovorans* strains. Acetamide hydrolysis, mediated by an aliphatic amidase, is a practical reaction for identifying *P. acidovorans* strains. This enzyme occurs also in pyocyanogenic, but few apyocyanogenic, *P. aeruginosa* strains and most *P. cepacia* strains.

Isolation and Relationship to Disease

P. acidovorans and *P. testosteroni* have been isolated from clinical specimens; intravenous tubing; urinary catheter; ultrasonic nebulizer; serum; feces of human, rabbit, turtle, frog, and cobra; sewer, ditch, river, well, and fish aquarium waters; raw milk; frozen shrimp; eye wash; and soil. They have a wide geographic distribution and have been isolated in the United States, Australia, England, Italy, Japan, India, France, the Netherlands, and Java. They are rarely opportunistic pathogens. *P. testosteroni* was associated with septicemia and conjunctivitis. *P. acidovorans* (along with *Enterobacter cloacae)* was the etiologic agent of septicemia in five patients undergoing cardiovascular monitoring.

DIMINUTA GROUP

The two species in the diminuta group have the characters of the genus *Pseudomonas* are recorded. The distinctive characteristic of the strains of these species is the very tightly coiled monotrichous flagellum, with a wavelength that varies from *0.62* to *0.98* gm. The wavelength of most polar monotrichous pseudomonads is approximately 2 μm. Specific growth factors are required; however, the species in the diminuta group grow in most peptone media without growth factor supplement. Indophenol oxidase is produced. Acid is produced from primary alcohols by all strains that can utilize alcohols, and some strains hydrolyze gelatin and starch. Members of the diminuta group in general have a restricted range of biochemical activities.

PSEUDOMONAS DIMINUTA AND PSEUDOMONAS VESICULARIS

Identification

The characteristics of 42 strains of *P. diminuta* and 32 strains of *P. vesicularis* are recorded. Minimal characters for identification of most *P. diminuta* and *P. vesicularis* strains are recorded. The flagellar morphologies of the type strains of *P. diminuta* and *P. vesicularis* have been described and illustrated.

Most carbohydrates in OF basal medium are not oxidized by *P. diminuta.* Pantothenate, biotin, cyanocobalamin, and cystine are required for growth. Carotenoid pigments are not produced. *P. diminuta* strains produce a pellicle in broth culture.

P. vesicularis strains may oxidize glucose, maltose, and other carbohydrates in OF basal medium; acid production, however, may be very weak, equivocal, or absent. The hydrolysis of esculin by *P. vesicularis is* a useful method for distinguishing this species from *P. diminuta.* Pantothenate, biotin, and cyanocobalamin are required for growth. Colonies of some strains produce orangered carotenoid pigments. A brown discoloration may accumulate in the agar medium surrounding colonies. *P. vesicularis* does not produce a pellicle in broth culture. *P. vesicularis* and most *P. diminuta* strains produce acid from ethanol.

Isolation and Relationship to Disease

P. diminuta has been isolated from stream and ditch waters, contaminated cell culture, respiratory equipment, and various patient sources. *P. vesicularis* has been isolated from the urinary bladder epithelium of a medicinal leech, whooping cough plate, tap water aerator, and clinical sources. *P. diminuta* was the etiologic agent in one case of septicemia.

PSEUDOMONAS MALTOPHILIA

Identification

P. maltophilia has the characters of the genus *Pseudomonas* are recorded. The characteristics of 527 strains of *P. maltophilia* are recorded. Minimal characters for identification of most *P. maltophilia* strains are recorded. The type strain has been described, and the name has been revived.

The morphological and biochemical reaction pattern of *P. maltophilia* is remarkably uniform. Colonies develop a characteristic lavender-green color on blood agar media. Brown-colored by-products of metabolism may accumulate in certain agar media. The -intensity of the brown color is enhanced by such factors as elevated incubation temperature. A similar reaction is observed among other glucose-nonfermenting gram-negative rods. Some strains slowly produce a very faint yellow, intracellular pigment which does not diffuse out of the colony into the agar medium. This water-soluble, yellow pigment may be due to flavins. Water-soluble melanin is not produced in tyrosine-containing media. Hemolysis does not occur around discrete colonies, but there is a greenish discoloration of erythrocytes around confluent growth. Growth on blood agar is accompanied by a strong odor of ammonia.

TABLE 6.14: CHARACTERS USED IN IDENTIFICATION OF DIMINUTA GROUP *PSEUDOMONAS* SPECIES

Test, substrate, or morphology	*P. diminuta (42 strains)*		*P. vesicularis (32 strains)*	
	Sign%	*Positive*	*Sign %*	*Positive*
Acid:Glucose, 1 %	–	0	– or +(w)	41
(OFBM)Fructose	–	0	–	0
Galactose	0	– or +(w)	–	19
Man nose	–	0	–	0
Rhamnose	0	– or +(w)	–	16
Xylose	0	– or +(w)	–	16
Lactose	–	0	–	0
Sucrose	–	0	–	0
Maltose	0	– or +(w)	–	47
Mannitol	–	0	–	0
Lactose, 10% (PAB)	–	0	–	0
ONPG	–	0	+ or –	50
Pyoverdin	–	0	–	0
Hydrogen sulfide	–	0	–	0
(KIA)Nitrate reduction	–	2	–	3
Gas from nitrate	–	0	–	0
Indophenol oxidase	+	100	+	100
Arginine dihydrolase	–	0	–	0
(DBM)Lysine decarboxylase	–	0	–	0
Ornithine decarbox0 ylase	–	0	–	0
Phenylalanine deam	–	7	–	0
inaseHydrolysis:Urea	–	0	–	0
Esculin	–	0	+	100
Tween 80	– or+	12	– or+	16
Starch	–	0	+	94
DNA	– or+	14	–	0

Lecithin	–	0	–	0
Gelatin	+ or –	62	+ or –	56
Acetamide	–	0	–	0
Hemolysis	–	0	–	0
Growth on:SS	–	0	–	0
MacConkey	+98	– or +	19	6.5%
NaCl	–	0	–	0
Cetrimide (PA)	–	0	–	0
MBM + acetate	–	0	–	0
Growth at 42°C	– or+	21	–	0
Polymyxin susceptible	+ or –	50	+ or –	88
Motile	+	100	+	97
No. of flagella		1		1

Some cells in all strains of *P. maltophilia* have a polar tuft of three or more flagella. Oxidative acidity accumulates in OF maltose and glucose media; how ever, the acid reaction is more pronounced in maltose than in glucose. OF glucose medium may remain neutral or become weakly acid after 18 to 24 h. An ONPG reaction is produced by most strains. *P. maltophilia* produces high levels of lactose dehydrogenase which, rather than beta-galactosidase, is probably responsible for the metabolism of lactose. Lysine decarboxylase, esculin, DNase, and gelatin reactions are produced. Although a weak indophenol oxidase reaction is produced by a few *P. maltophilia* strains, the species is usually described as negative for this test. *P. maltophilia, P. syringae,* and other phytopathogenie, fluorescent pseudomonads appear to lack cytochrome c, which may be necessary for the indophenol oxidase reaction.

Most, but not all, strains of *P. maltophilia* require either methionine or cystine plus glycine for growth: however, peptone-containing media, without growth factor supplement, support growth of all strains. Strains that do not require growth factors (biovar 2) differ from methionine-requiring strains (biovar 1) by the manner in which they assimilate select carbon compounds and other properties.

Proposals have been made to transfer *P. maltophilia* to the genus *Xanthomonas* as a separate species, *Xanthomonas maltophilia*. This is based partly on the following common reactions: acid reactions from OF glucose, lactose, and maltose; alkaline reactions from rhamnose and mannitol; negative indophenol oxidase reaction;

hydrolysis of esculin and ONPG; and a requirement for amino acids as growth factors.

Isolation and Relationship to Disease

P. maltophilia is a free-living, ubiquitous organism with a wide geographic distribution. It has been found in raw and pasteurized milk; distilled,. well, stagnant, and river waters; sewage; external ear canals and skin of diver; spray atomizer;p hard lens cleaning solution; frozen fish; feces of snakes, lizards, frogs, and rabbits; rotten eggs; and soil in petroleum zones. It has been recovered from the internal tissue of banana pseudostem, decaying banana sucker, cotton seed, bean pod, tobacco seedlings, and rhizosphere of cabbage and other cultured plants. In the hospital environment, it has been isolated from contaminated cell culture, streptomycin solution, contaminated deionized water, sink drains, water bath, faucet aerator, incubator reservoirs, respirators, nebulizers, intravenous catheters, evacuated blood collection tubes, blood monitoring transducers, and cardiopulmonary bypass equipment.

P. maltophilia is usually a commensal or contaminant and is part of the transient flora of hospitalized patients. It has been recovered from almost every site on or within the body. Since *1967,* this organism has been associated with primary pneumonia and opportunistic infections including endocarditis, septicemia, broncho-, lobar, and aspiration pneumonia, urinary tract infections, cholangitis, conjunctivitis, acute mastoiditis, meningitis, wound infections, and abscesses. Enzymes including esterases, lipases, mucinase, hyaluronidase, RNase, and hemolysins may be potential virulence factors.

PSEUDOMONAS MESOPHILICA

Identification

P. mesophilica has the characters of the genus *Pseudomonas* are recorded. The characteristics of six strains of *P. mesophilica* are recorded. Minimal characters for identification of most *P. mesophilica* strains are recorded. The type strain has been described.

Colonies are shiny, smooth, round, raised, entire, and rarely visible before 3 days of incubation at 30°C. An intracellular, nondiffusible, water-insoluble, pink, oxocarotenoid pigment is produced. Pink pigmentation is a characteristic of most methanol-utilizing bacteria. Acid is produced slowly, within 2 to 4 days, from fructose, xylose, and methanol in OF basal medium. Indophenol oxidase, urease, and amylase reactions are produced. Strains are motile, with polar monotrichous flagella. They have *65.8* mol% G+C content in their DNA.

Isolation and Relationship to Disease

P. mesophilica has been isolated from tobacco leaves; perennial rye grass leaves; methane-oxidizing enrichment cultures from soil, sewage, and rumen of a cow; water; nebulizer; and human clinical materials.

PSEUDOMONAS PAUCIMOBILIS

Identification

P. paucimobilis has the characters of the genus Pseudomonas are recorded. The characteristics of 110 strains of *P. paucimobilis* are recorded. Minimal characters for identification of most *P. paucimobilis* strains are recorded. The type strain has been described.

Colonies are either butyrous, or mucoid and viscid, and develop an intracellular, nondiffusible, carotenoid (nostoxanthin) pigment on most media. This yellow pigment is distinct from the brominated arylpolyenes (xanthomonadins) present in Xanthomonas species. No brown pigments are produced. The name *paucimobilis* indicates that only a few cells of a strain in a population may be motile. Motility is best demonstrated at room temperature. When motile, they are polar monotrichous. Indophenol oxidase is produced, and ONPG and esculin are hydrolyzed. Acid is produced in OF basal medium from a number of carbohydrates but not from sugar alcohols. *P. paucimobilis* grows in mineral base medium with ammonium ion as sole source of nitrogen and acetate as sole source of carbon and energy. A negative test for urease distinguishes this species from phenotypically similar strains of *Sphingobacterium (Flavobacterium) multivorum.*

Evidence from DNA hybridization results, rRNA cistron comparisons, and cellular fatty acid profiles indicates that *P. paucimobilis is* not an authentic member of the genus *Pseudomonas* and that a new genus designation may be justified.

Isolation and Relationship to Disease

P. paucimobilis has been isolated from clinical specimens; respirators; humidifiers; ultrasonic nebulizers; infusate and dialysis fluids; water bath; tap water aerator; incubator, tap, distilled, boiled, bottled mineral, and swimming pool waters; blow bottles; water bottles; contaminated cell culture; and surgical unit air. *P. paucimobilis* has been implicated in an infected leg ulcer after trauma, postoperative chronic cellulitis, septicemia, acute meningitis, and urinary tract infection. DNase, esterases, phosphatases, lipase, and endotoxin may contribute to its virulence.

PSEUDOMONAS PUTREFACIENS

Identification

P. putrefaciens has the characters of the genus *Pseudomonas* are recorded. The characteristics of 72 strains of *P. putrefaciens* are recorded. Minimal characters for identification of most *P. putrefaciens* strains are recorded. (ATCC 8071 was designated) and acknowledged as the "type strain" of *Alteromonas (Pseudomonas) putrefaciens.*

Cells of *P. putrefaciens* strains possess polar monotrichous flagella. Some strains may contain cells with lateral flagella. Colonies on agar media are slightly viscous or mucoid and usually reddish-brown or pink.

P. putrefaciens is the only glucose-nonfermenting gram-negative rod known to produce hydrogen sulfide in KIA and TSIA. For this reason strains have been mistaken for rare anaerogenic glucose-fermenting isolates of *Salmonella* and *Proteus* species. Occasionally, strains produce only a small quantity of or no hydrogen sulfide in the butts of KIA or TSIA.

The production of acid from glucose is oxygen dependent and may be weak and delayed for several days. Ornithine decarboxylase and DNase tests are positive. Most strains grow at 30°C, many fail to grow at 35°C, and some are psychrophilic. Growth usually develops in mineral base medium containing ammonium ion as sole source of nitrogen and acetate as sole source of carbon and energy. Sodium cations are required for growth of some strains.

Two distinct groups of strains exist within *P. putrefaciens,* and additional groups have been described. Biovar 1 (CDC group Ib1) has a relatively low G+C content of 49 mol% in its DNA and consists primarily of strains from environmental sources. These strains produce acid from OF sucrose and maltose and fail to grow in 6.5% NaCI or on salmonella-shigella agar. Biovar 2 (CDC group Ib-2), with a high G+C content of 58 mol% in its DNA, includes mostly strains from human clinical specimens. These strains grow in 6.5% NaC1 and on salmonella-shigella agar but do not produce acid from disaccharides.

Bacteria of low G+C content have been excluded from the genus *Pseudomonas* and transferred to the new genus Alteromonas created precisely to accommodate Pseudomonas-like bacteria of low G+C content. The transfer of this group of *P. putrefaciens* strains to the genus *Alteromonas* as *Alteromonas putrefaciens* has not found uniform support.

Isolation and Relationship to Disease

P. putrefaciens is found in soil and water and has a wide geographic distribution. It has been found in raw and sweet milk and cream; eggs; snake feces; sewage; stagnant, sea, well, lake, and river waters; contaminated cell culture; external ear canals of divers; oil emulsion; and natural gas and petroleum brines. It is responsible for putrid deterioration of butter, hydrogen sulfide spoilage in haddock fillets, and green discoloration of fresh meat and vacuum-packed beef, and it has been isolated from fresh, frozen, and stored cod fillets, ground beef, and frozen poultry carcasses.

Isolates of *P. putrefaciens* from clinical specimens of human origin in most instances represent colonization in the absence of clinical disease. *P. putrefaciens* has been etiologically associated with septicemia, an intraabdominal abscess, otitis media, an infected tibia and overlying ulcer after trauma, and purulent skin ulcerations of the legs of diabetics and debilitated patients. Intraperitoneal injection of cells of *P. putrefaciens* in mice resulted in 50% lethal dose values of 1.2×10^9 cells, indicating that death was more likely due to endotoxin and not to invasiveness.

VE GROUP OF PSEUDOMONAS SPECIES

The species in the Ve group have the characters of the genus Pseudomonas are recorded. Strains in the Ve group may produce smooth colonies or wrinkled, rough, and adherent as well as smooth colonies. The colonies produce an intracellular, nondiffusible, water-insoluble, yellow pigment. Aggregates of cells in sausage-shape chain formation (symplasmata) occur in rough colony forms only. Symplasmata are seen with the low-power objective of a microscope in a hangingdrop preparation of a 24-h broth culture or condensate from an agar slant culture.

Acid is produced from a number of carbohydrates and polyhydric alcohols in OF basal medium. These species grow in mineral base medium containing ammonium ion as sole source of nitrogen and glucose as sole source of carbon and energy. Ve group strains are motile, use polyhydric alcohols, and do not produce indophenol oxidase. These characteristics distinguish Ve from *P. paucimobilis* strains. Two subdivisions of Ve group strains are recognized.

Pseudomonas sP. Ve-1 and Pseudomonas sP. Ve-2 Identification

The characteristics of 21 strains of *Pseudomonas sP.* Ve-1 and 77 strains of *Pseudomonas sP.* Ve-2 are recorded. Minimal characters for identification of most *Pseudomonas spP.* Ve-1 and Ve2 strains are recorded.

Pseudomonas sP. Ve-1 strains have a polar tuft of flagella and hydrolyze esculin. *Pseudomonas sP.* Ve-2 strains are polar monotrichous and do not hydrolyze esculin. Strains of the former may produce arginine dihydrolase and nitratase. Cellular fatty acid composition shows group Ve strains to be similar to other species of *Pseudomonas. Pseudomonas sP.* Ve-1 strains have 56.8 mol% G+C content in their DNA, which is at the lower limit of the range for *Pseudomonas* species. The 66.9 mol% G+C content in the DNA of *Pseudomonas sP.* Ve-2 strains is near the G+C values of *Chromobacterium lividum, Chromobacterium violaceum,* and many *Pseudomonas* species. Each warrants species recognition.

Isolation and Relationship to Disease

Pseudomonas spP. Ve-1 and Ve-2 have been isolated from a number of clinical specimens. In the hospital environment, *Pseudomonas sP.* Ve-2 has been recovered from respiratory therapy equipment and sink drains. Bacteremia caused by *Pseudomonas sP.* Ve-1 occurred in a patient after skull fracture and surgery. An immunocompromised patient developed septicemia attributed to *Pseudomonas sP.* Ve-1.

OTHER *PSEUDOMONAS* SPECIES

Pseudomonas pertucinogena, identified as rough phase IV *Bordetella pertussis,* has the characters of the genus *Pseudomonas.* The organism is polar monotrichous and produces indophenol oxidase, phenylalanine deaminase, and pertucin, a bacteriocin that inhibits the growth of *B. pertussis.*

Recently described *Pseudomonas-like* group 2 strains have features similar to species in the pseudomallei group but are distinguished from the latter by cellular fatty acid composition and select conventional biochemical tests.

7

Sexually Transmitted Diseases

Adolescents are at higher risk for sexually transmitted infection than other age groups for a variety of behavioral, biological, and psychosocial reasons. There are a number of factors which specifically influence sexually transmitted disease (STD) risks in this group. Foremost among these factors is increasing sexual activity. For example, the percentage of 15- to 19-yearold females who were sexually active increased from 30 to 50% from 1971 to 1979. Studies in adolescent males showed parallel increases. In another study, 7.6% of white and 15.3% of black adolescents had sexual relations before 13 years of age and by age 20 these rates rose to 65 and 85%, respectively. Other risk factors for adolescents include failure to use any form of contraceptive and increased use of crack and other illicit substances.

The most common STDs are those caused by *Chlamydia trachomatis* and human papillomavirus (HPV) infections. Teenage females appear to have an increased risk of chlamydial infections. The prevalence of genital *C. trachomatis* infections in female adolescents ranges from 4.9% in college students to 27% in predominantly urban, lower socioeconomic-class pregnant adolescents. In addition, 8-40% of asymptomatic young women who have cervical cultures are found to be infected with *C. trachomatis*. HPV, which causes genital warts and cervical neoplasia, has been found in 3-8% of unselected young women and in 25% of STD clinic attenders. Although the rate of gonorrhea amongst adolescents aged 15-19 fell to 8.8% between 1981 and 1988, the rates for those aged 10-14 actually increased over the same period. The incidence of syphilis is estimated to be rising at a rate of 100,000 new cases annually, with a disproportionate increase among blacks. Although teenagers have suffered less from this epidemic, the rate of syphilis among minority teenage women has been increasing. The incidence of chancroid rose from 665 in 1984 to 4,714 in 1989. The disease is usually more prevalent in men than in women and uncircumcised males, A further concern is that chancroid may facilitate human immunodeficiency virus (HIV) transmission. Trends in genital herpes simplex virus (HSV) infections among teenagers have followed the same course as their bacterial counterparts: approximately 4% of whites and 17% of blacks have been infected with HSV-2 by the end of their teenage years.

The morbidity associated with these STDs is considerable. Gonococcal or chlamydial pelvic inflammatory disease (PID) is a major health problem for adolescents. Recently, HPV infection has been implicated in genital cancers. The following discussion will focus on the common pathogens responsible for STDs in adolescents with specific emphasis on treatment and prevention.

CHLAMYDIA

Chlamydiae are obligate intracellular bacteria which require cultured cells for isolation. The two major modes of transmission of *C. trachomatis* are sexual and congenital.

Clinical Manifestations

Chlamydia appears to be at least two times more common in adolescents than gonorrhea. Risk factors for genital chlamydial infections in women include young age, unmarried status, multiple sexual partners or recent partner changes, pregnancy, urban living and use of the oral contraceptives.

Chlamydial cervicitis is common and its clinical recognition depends upon a high index of suspicion and a careful cervical examination. In symptomatic cases, the most common complaints are discharge (90%), symptoms of urethritis (20%), intermenstrual spotting (10-15%) or sudden hypermenorrhea. Findings on physical examination suggestive of chlamydial infection include easily induced endocervical bleeding, mucopurulent endocervical discharge and edema of the area of ectopy.

The most important complication of chlamydia infections in female patients is acute salpingitis. PID results when the infection spreads to the fallopian tubes, ovaries and contiguous structures. Eight to 10% of teenage girls develop PID following chlamydia cervicitis. Several factors predisposing adolescents to PID include the presence of cervical ectopy, relative lack of protecting neutralizing antibodies, anovulatory cycles, lack of contraceptive use, poor motivation for diagnosis and treatment, and poor compliance for follow-up. The risk decreases with increasing age. *C. trachomatis* is responsible for up to 38% of inpatients and 52% of outpatients treated for PID.

The clinical manifestations associated with PID include lower abdominal pain, purulent cervical discharge, cervical motion tenderness, adnexal tenderness, and fever. The erythrocyte sedimentation rate and C-reactive protein (CRP) are elevated in up to 75% of cases. Laparoscopy will confirm about 65% of clinically diagnosed cases of PID. The clinical course of chlamydial PID is less dramatic than that of gonococcal PID. More than 50% of patients with tubal infertility have no history of PID, suggesting that subclinical infection is frequent.

The most important complication of PID is permanent tubal dysfunction and infertility. Infertility rates of 13% after single episodes of salpingitis, 36% after two episodes, and 75% after three or more episodes have been reported. Women with chlamydial salpingitis are also at increased risk of ectopic pregnancy. Complications arising from the intraabdominal spread of PID include periappendicitis, perisplenitis and perihepatitis (FitzHugh-Curtis syndrome). In men, the most common manifestation of chlamydia infection is urethritis. Rates of isolation of *C. trachomatis* from urethral specimens vary from 8 to 1 A in asymptomatic males and are up to 35% in symptomatic males of lower socioeconomic status.

The time between sexual intercourse with an infected partner and development of symptoms of urethritis varies from 7 to 14 days. Initial symptoms include discharge and/or dysuria. Urethral examination reveals an inflamed meatus. This urethral discharge may be clear or mucopurulent, scant or copious.

The most serious complication of chlamydial infections in males is epididymitis. However, there are no data on the risk of epididymitis in male adolescents. Approximately 1-3% of men with symptomatic nongonococcal urethritis (NGU) develop epididymitis. Epididymitis usually presents with unilateral scrotal swelling, tenderness and fever and is often accompanied by symptoms of urethritis. Epididymitis caused by chlamydia has been associated with oligospermia and azospermia.

Chlamydial proctitis, which may occur in homosexual men who practice passive rectal intercourse, is difficult to clinically differentiate from proctitis of other causes. Chlamydia-related arthritis, traditionally known at Reiter's syndrome, had been termed sexually acquired reactive arthritis or SARA. It is estimated that approximately 1% of men with NGU develop reactive arthritis.

Laboratory Diagnosis

C. trachomatis can be detected in clinical specimens by culture, antigen detection using monoclonal antibodies, enzyme-linked immunosorbent assay, direct or indirect immunofluorescent assays or a combination of these. DNA hybridization and DNA amplification using polymerase chain reaction (PCR) have recently been reported from several research laboratories.

The prevalence of chlamydia infection in almost all populations of male adolescents is high enough to warrant screening their urine with dipsticks. *C. trachomatis* has been isolated from 69 to 85% of asymptomatic sexually active boys with pyuria of greater than or equal to 10 leukocytes per highpower field. Confirmation of the diagnosis can be made by the use of culture or rapid diagnostic tests. Since *C. trachomatis is* an obligate intracellular parasite, it is imperative that infected cells be obtained when obtaining clinical specimens. In men, the best specimen is obtained by rotating a swab inserted several centimeters into the urethra. Endocervical cells must be collected as *C. trachomatis* does not infect squamous cells. In the adolescent female the transitional zone is usually easily exposed making specimen collection easier. Wooden sticks and calcium-alginated tipped swabs should not be used to obtain cultures because they are toxic to the organism. Specimens obtained using brushes and scrapings are more sensitive than specimens obtained with swabs.

Specimens should be placed directly into a medium containing antibiotics (vancomycin, gentamicin) and antifungal agents (amphotericin B) and transported promptly to the laboratory. McCoy cells are the most sensitive culture system. Direct fluorescent antibody (DFA) and enzyme immunoassay procedures provide useful alternatives to culture. The sensitivities and specificities among these antigen detection techniques range from 85 to 93% and 94 to 99%, respectively.

Serologic tests are available but have not found application in routine diagnosis of chlamydial genital tract infections because most patients attending STD clinics have had previous exposure to *C. trachomatis* and will be seropositive.

TABLE 7.1 : CDC GUIDELINES FOR TREATMENT OF *C. TRACHOMATIS* INFECTIONS

Uncomplicated urethral, endocervical or rectal C. Trachomatis *infections*	
Recommended	Doxycycline 100 mg p.o. b.i.d. × 7 days, or Tetracycline 500 mg p.o. q.i.d. × 7 days
Alternative	Erythromycin base 500 mg p.o. q.i.d. × 7 days If erythromycin not tolerated, sulfisoxazole 500 mg p.o. q.i.d. × 10 days
C. trachomatis *infection in pregnancy*	
Recommended	Erythromycin base 500 mg p.o. q.i.d. × 7 days. If not tolerated,
Alternative	Erythromycin base 250 mg p.o. q.i.d. × 14 days, or Erythromycin ethylsuccinate 800 mg p.o. q.i.d. × 7 days, or Erythromycin ethylsuccinate 400 mg p.o. q.i.d. × 14 days. If not tolerated, Amoxicillin 500 mg p.o. t.i.d. × 7 days (limited data on efficacy) *Erythromycin estolate is contraindicated in pregnancy because of hepatotoxicity*
Nongonococcal urethritis	
Recommended	Doxycycline 100 mg p.o. b.i.d. × 7 days, or Tetracycline 500 mg p.o. q.i.d. × 7 days
Alternative	Erythromycin base 500 mg p.o. q.i.d. × 7 days Erythromycin ethylsuccinate 800 mg p.o. q.i.d. × 7 days. If not tolerated, Erythromycin ethylsuccinate 400 mg p.o. q.i.d. × 14 days, or Erythromycin base 250 mg p.o. q,i.d. × 14 days
Recurrent nongonococcal urethritis (consider noncompliance or reinfection): repeat treatment	
Recommended	Doxycycline 100 mg p.o. b.i.d. × 7 days, or

	Tetracycline 500 mg p.o. q.i.d. × 7 days
Alternative	Erythromycin base 500 mg p.o. q.i.d. × 7 days, or Erythromycin base 800 mg p.o. q.i.d. × 7 days. If not tolerated.
	Erythromycin base 400 mg p.o. q.i.d. × 14 days, or
	Erythromycin base 250 mg p.o. q.i.d. × 14 days
Mucopurulent cervicitis	
Recommended	If *N. gonorrhoeae* found on- gram stain or culture, treat as recommended for uncomplicated gonorrhea, including co-treatment for *C. trachomatis;* if *N. gonorrhoeae is* not found, treatment is the same as for nongonococcal urethritis
Epididymitis	
Recommended	Ceftriaxone 250 mg i.m. once, and
	Doxycycline 100 mg p.o. b.i.d. × 10 days, or
	Tetracycline 500 mg p.o. q.i.d. × 10 days

Treatment

The 1989 Center for Disease Control (CDC) guidelines for treating *C. trachomatis* infections are provided in table 1. Doxycycline is the preferred agent as dosing is less frequent, its absorption is not impaired by the simultaneous ingestion of food or milk, and it is not photosensitizing. The CDC recommends that the sexual partners of patients with *C. trachomatis* infection undergo testing and receive treatment if the contact occurred within 30 days of the onset of symptoms. In pregnant adolescents, nursing mothers, infants, children, and those unable to tolerate the tetracyclines, erythromycin is the drug of choice. This may be associated with gastrointestinal disturbance.

Hospitalization is recommended for all adolescents with PID because compliance with therapy is unpredictable and the long-term sequelae of PID may be very serious.

A test of cure is not currently recommended by the CDC since antibiotic resistance has not been identified. However, since *C. trachomatis* infections are associated with significant morbidity follow-up cultures after treatment may be prudent.

CHANCROID

Chancroid is caused by *Haemophilus ducreyi,* a gram-negative, facultative anaerobic bacillus. There is very little literature concerning chancroid specifically in the adolescent population.

TABLE 5.2. TREATMENT REGIMENS FOR ACUTE PID

Recommended inpatient regimens

1. Cefoxitin: 2.0 g i.v. q6h or cefotetan[i]: 2.0 g i.v. gl2h plus
   Doxycycline: 100 mg i.v. g12h
   A. Use i.v. for minimum of 48 h after patient improves
   B. After discharge continue doxycycline 100 mg p.o. b.i.d. to complete 10- to 14day course

   or
2. Clindamycin: 900 mg ix. q8h
   plus
   Gentamicin: 2.0 mg/kg i.v. or i.m. in one loading dose, followed by a maintenance dose of 1.5 mg/kg i.v. q8h in patients with normal renal function
   A Use i.v. for minimum of 48 h after patient improves
   A. Use i.v. for minimum of 48 h after patient improves
   B. After discharge, doxycycline 100 mg p.o. b.i.d. for 10-14 days total

   Continue clindamycin 450 mg p.o. q.i.d. to complete a 10- to 14-day course as an alternative

Recommended alternative for patients who do not tolerate doxycycline

1. Erythromycin: 500 mg p.o. q.i.d. x 10-14 days (based on limited clinical data)

Management of sex partners

Partners should be evaluated for STDs and treated empirically for *N. gonorrhoeae* and *C. trachomatis* infection

---

Clinical Manifestations

Chancroid is an acute ulcerative disease of the external genitalia which occurs after an incubation period of 3-10 days. The first sign of infection is a small hyperemic macule. The macule evolves into a papule, pustule and finally a small painful ulcer. Adjacent lesions may coalesce and form a giant ulcer. Patients who are left untreated or who are not treated appropriately may suffer extensive ulceration and tissue destruction with massive edema of the vulva and perineum and suppurative buboes with draining fistulas. In adolescent females, the primary lesion is usually located on the labia minora although it may also be found on the vulva, the vaginal mucosa, the cervix, or the clitoris. The infected female may present with pain on urination and defecation, dyspareunia, or increased vaginal discharge. In the male adolescent the lesions may be on either the external or internal surface of the prepuce, the frenulum, or in the coronal sulcus. The glans meatus and the shaft of the penis are less often involved.

Chancroid can infect contiguous tissues such as the cervix, the upper part of the thigh, posterior aspects of the buttocks, and perianal areas by autoinoculation.

Painful inguinal adenitis is a characteristic feature of chancroid and is present in up to 50% of patients. The adenitis is unilateral in most patients and there is typically erythema of the overlying skin.

Diagnosis of Chancroid

Direct microscopic examination of stained smears collected by scrapings of the base and edges of the ulcer is the best initial diagnostic test. The presence of *H. ducreyi* can be established presumptively by noting the presence of gram-negative coccobacilli in a school-of-fish or railroad-track pattern on Gram's stain. The presence of *H. ducreyi* is confirmed by culture utilizing laboratory conditions similar to those required for *Neisseria gonorrhoeae*. DNA probes for the identification of *H. ducreyi* are under investigation and no reliable serologic test is available. Since one-third of patients with chancroid have more than one etiology, other causes of STDs should be pursued.

Treatment

The recommended treatment for chancroid is oral erythromycin, 500 mg 4 times a day for 7 days, or a single intramuscular injection of 250 mg of ceftriaxone. The alternative therapies of trimethoprim-sulfamethoxazole, 160/800 mg 2 times a day for 7 days, and amoxicil-linclavulanic acid, 500/125 mg 3 times a day for 7 days, also appear to be effective, although there has been little experience with these drugs in the United States. Control of outbreaks of chancroid requires prompt identification and treatment of sexual contacts, whether they are symptomatic or asymptomatic.

Chancroid is a major cofactor in the heterosexual transmission ofhuman immunodeficiency virus (HIV) and as a result simple, effective, and inexpensive treatment regimens are needed for HIV control programs. Single-dose trcatmat with quinolone (fleroxacin) was found to be less effective in HIVpositive patients with chancroid than in HIV-negative patients.

GONORRHEA

Neisseria gonorrhoeae are aerobic nonmotile and non-spore-forming gram-negative diplococci.

Clinical Manifestations

Patients with gonococcal cervicitis have nonspecific symptoms including vaginal discharge, dysuria, urinary frequency, and dyspareunia. Abnormal cervical mucus with cervical edema and friability is evident on speculum examination. Labial swelling and pain occur if Bartholin's gland infection is also present. Gonococcal urethritis and cervicitis can coexist in female adolescents. The urethral infection may be asymptomatic or the patient may present with urinary frequency, internal dysuria, suprapubic pain, and purulent discharge from the urethra or Skene's glands.

Symptomatic patients with urethritis have purulent urethral discharge accompanied by dysuria. Without treatment, signs and symptoms of urethritis typically peak within 2 weeks, during which time complications of posterior urethritis and epididymitis are most likely to occur. Asymptomatic infection accounts for approximately 5% of cases of gonococcal urethritis.

It is often difficult to distinguish gonococcal urethritis from NGU caused by *C. trachomatis*. Although gonococcal urethritis is more frequently associated with purulent discharge and dysuria, co-infection with *C. trachomatis* is very common.

Rectal infections are common in women and homosexual men. About 40% of women with gonococcal cervicitis and urethritis will have concomitant isolation of the rectum. Rectal gonorrhea occurs in men who practice receptive rectal intercourse, whereas rectal infection in women is most commonly due to perianal contamination with cervicovaginal secre*tions. Most patients with anorectal gonorrhea are asymptomatic. Symptoms,* when present, include tenesmus, mucoid discharge, hematochezia, perianal irritation, and constipation. Anorectal specimens may be obtained by inserting the swab 2-3 cm into the anal canal and rotating it against the rectal walls. Additional anorectal specimens from symptomatic patients should be obtained under direct vision using anoscopy as the sensitivity of the smear is increased.

Oropharyngeal gonorrhea results from orogenital contact and is more efficiently acquired by fellatio than by cunnilingus. Exudative pharyngitis and sore throat are seen in symptomatic cases. However, over 90% of pharyngeal infections are asymptomatic.

PID is the most common complication of gonorrhea in women. It is estimated that 10-20% of untreated or undertreated women with gonococcal cervicitis develop PID. Patients at highest risk of developing PID are unmarried adolescents with a history of multiple sexual partners, early age at first intercourse, and past history of STD.

Patients with gonococcal salpingitis usually present with crampy, dull abdominal pain, dyspareunia, abnormal menses or intermenstrual bleeding. There is tenderness on bimanual examination, abnormal cervical or vaginal discharge, and sometimes an adnexal mass or tuboovarian abscess. Gramnegative intracellular diplococci may be seen on gram-stained smears of cervical secretions. However, the gram stain is negative in up to 40-60% of infected women. When compared with women with nongonococcal salpingitis, women with gonococcal salpingitis are more likely to be febrile and to develop symptoms earlier in the course of infection.

Disseminated gonococcal infection (DGI) occurs in 0.5-3% of patients with untreated mucosal gonorrhea. Women outnumber men by three to one, and pregnant women have the highest risk. Only 20-30% of DGI patients have positive blood cultures. This incidence decreases as the duration of clinical signs and symptoms increases.

Two forms of gonococcal arthritis have been described. The first group includes patients with reactive migratory polyarthralgias and tenosynovitis.

These patients typically present with fever, chills, and a rash. The second common syndrome is characterized by a monoarticular effusion, affecting large joints, most commonly the knee. Systemic symptoms are usually absent and positive synovial fluid cultures occur in 20-50% of cases.

Gonococcal endocarditis, meningitis and toxic hepatitis have been reported.

Diagnosis

Symptomatic gonococcal urethritis in men is most efficiently diagnosed by staining a smear of urethral exudate. This has a sensitivity of 90-95% and a specificity of 95-100%. Culture in women remains the standard for the diagnosis of gonorrhea and stained smears should be used as an adjunct to but not a replacement for culture. Swabs for culture must be taken promptly to the laboratory in an appropriate transport media or plated directly at the bedside. Gram's stain of smears obtained from the endocervix and rectum, if obtained under anoscopic guidance, are less sensitive (50-70% and 70-80%, respectively) but highly specific (both 95-100%). Blind-stained smears of anorectal specimens are too nonspecific for routine use in diagnosing infections at these sites.

Serologic tests have little diagnostic value in the diagnosis of gonococcal infections as they are neither sensitive nor specific enough for clinical use.

Treatment

The development of penicillin was a major landmark in the treatment of gonorrhea. However, surveillance data from the CDC estimate that over 7% of current isolates of *N. gonorrhoeae* are resistant to penicillin and tetracycline resistance is becoming more common. As a result of these emerging patterns of resistance, the CDC's 1989 STD Treatment Guidelines recommend that all gonococcal infections, including local and disseminated infections, be treated presumptively as if they were caused by antibiotic-resistant organisms.

Ceftriaxone is highly effective even against penicillin and tetracycline-resistant strains, can be given in a single small-volume injection in the deltoid, and is also likely to be effective against incubating syphilis. Intramuscular spectinomycin and oral ciprofloxacin or ofloxacin are single-dose alternatives for penicillin- and cephalosporin-allergic patients, but none of these agents is effective against incubating syphilis. All patients treated for gonorrhea should also be co-treated for chlamydial infections.

TABLE 7.3. CURRENT REGIMENS FOR THERAPY OF UNCOMPLICATED GONORRHEA.

1. Ceftriaxone, 250 mg, intramuscularly, one dose (or equivalent doses of other thirdgeneration cephalosporin, such as cefuroxine axetil, 1 g p.o. with 1 g probenecid; or cefotaxime I g i.m.; or ceftizoxime 500 mg i.m.)
2. Ciprofloxacin, 500 mg p.o. in a single dose, or norfloxacin 800 mg p.o., single dose (should not be given to pregnant women or children under 16)

   In addition, all patients treated for gonorrhea should also be co-treated for chlamydial infection with oral
   - doxycycline, 100 mg b.i.d. for 7 days
   - or tetracycline, 500 mg q.i.d. for 7 days
   - or erythromycin base or stearate, 500 mg q.i.d. for 7 days

Repeat cultures for test of cure are recommended for all patients with gonorrhea. All known infected sites should be recultured. Sexual contacts should be treated for gonorrhea and evaluated for other STD. In addition, all patients with gonorrhea should have a serologic test for syphilis and should be offered confidential counseling and testing for HIV infection.

GENITAL HUMAN PAPILLOMAVIRUS

Human papillomaviruses (HPV) have closed, circular, double-stranded DNA and icosahedral protein envelopes. Papillomaviruses are classified into types and subtypes based on the amount of DNA homology. Based upon DNA analysis, over 60 subtypes of HPV have been identified.

Clinical Manifestations

Most genital warts are seen in young adults between the ages of 16 and 25 years. Several studies have demonstrated that HPV may be found in the genital tracts of 13-38% of inner-city and suburban adolescents. Risk factors for infection are similar to those for other STD and include age less than 25 years, multiple sexual partners, lack of condom use, previous STD, sex with an infected partner, pregnancy, and an altered immune state. The interval from exposure to detection of condylomata ranges from 3 weeks to 8 months.

The sexually active male adolescent represents a huge reservoir of undetected HPV infection because of the infrequency of routine genital examinations in this population. Lesions are usually localized to the shaft, prepuce, frenulum, corona and glans of the penis. The meatus, anus and scrotum also may be involved. Lesions are usually multiple and vary between 1 and 5 mm in diameter. Urethral involvement manifests with discharge or bleeding. Urethral condyloma may represent a relatively common cause of asymptomatic hematuria in males. Subclinical papillomavirus infections (SPI) or condyloma planum are invisible to the naked eye; they can occur anywhere on the penis, the perianal area, scrotum, and urethra. They are usually multiple, and may be confluent.

In females, condylomata acuminata usually first appears on the fourchette and adjacent labia. Infection may spread quite rapidly to the clitoris, perineum, vagina, cervix, anus and rectum. SPI of the vulva is only visible by colposcopy after the application of acetic acid. These areas may develop fissures and cause vulvodynia and dyspareunia.

There is evidence that HPV plays a major role in the causation of human anogenital cancers. The types most commonly found with cancer are 16, 18, 31, 33 and 35, those most frequent in benign lesions are 6 and 11. Another potential consequence of HPV infection is perinatal transfer of the virus from an infected mother to her offspring, resulting in laryngeal papillomatosis in the child.

Perianal warts are usually condylomata acuminata, which may be of substantial size. Anal warts are more common in males who have engaged in anal intercourse. Females with anal warts often report no history of anal sex, suggesting that autoinoculation may occur.

Diagnosis

Condylomata acuminatum may be grossly evident on examination of the anogenital area. Microscopic examination following the application of acetic acid may identify areas of SPI. This procedure involves soaking the skin of the genital area in vinegar or 3% acetic acid for 5 min followed by examination of the skin. A shiny white appearance of the skin that follows the application of acetic acid is called acetowhitening and represents foci of epithelial hyperplasia. Since most HPV infections are subclinical, other methods including cytology, colposcopy, and biopsy are often needed to facilitate diagnosis. Cytology is the most common method used for the detection of cervical disease. Although the Pap smear has poor sensitivity and specificity, it is useful as a mass screening method. The detection of koilocytes on routine Pap smear suggests HPV infection of the cervix. Such screening is useful in determining who should undergo colposcopy and biopsy. Unfortunately, there is no cell culture system for detecting HPV, so definitive diagnosis is based upon the detection of DNA, usually performed by Southern blot hybridization.

Treatment

No form of treatment has been shown to eradicate the virus. Podophyllin remains the most widely used treatment. The solution is applied to the warts with a sterile swab, allowed to dry, and washed off 4 h after the application. Applications are repeated weekly for up to 6 weeks. Because the concentration of podophyllin in commercially available solutions varies widely, treatment success rates range from 30 to 80%. Podophyllin causes both local and systemic reactions and its use is contraindicated in pregnancy and on vaginal, cervical and rectal mucosal lesions.

Trichloroacetic acid (TCA) is an alternative chemical therapy. It is provided in strengths of 25-85%. The solution should be applied directly to the lesion, avoiding application to normal skin. It can be used every 7-12 days for up to 6 weeks. The success rate is 20-30% after a single application of the 85% solution. Weekly treatments result in eradication rates slightly inferior to those of podophyllin. TCA is relatively safe, although local sensitivity reactions do occur. TCA can be used on mucosal surfaces and in pregnant women.

Liquid nitrogen is the usual cryogen used to treat condylomatous lesions. It is applied to lesions with a cotton or dacron swab for 5-10 s or until the entire lesion and/or surrounding 5 mm becomes white. Treatment should be repeated every 7-12 days, and if no response occurs after four to six applications, the patient should be referred for ablative therapy. Cryotherapy is nontoxic and is not associated with significant adverse reactions, other than local pain. It has been used to treat condylomata acuminatum in pregnant women.

5-Fluorouracil (5-FU) is a cytotoxic agent that causes necrosis and sloughing of growing tissue. 5-FU should be applied to the affected area(s) for 2 consecutive days once a week for 10 weeks. Severe inflammatory reactions, burns and scarring can occur.

Extensive, multifocal, or resistant genital warts should be treated with laser therapy. Local or general anesthesia is required for most laser treatments of genital HPV lesions. In addition to pain, complications may include vaginal discharge, periurethral swelling, and vulvar itching and swelling. Recurrence rates in various series have been 25% or higher. Prophylactic use of 5-FU following laser therapy reduces the likelihood of these recurrences.

Interferons are biologically active glycoproteins with antiviral, antiproliferative, and immunomodulatory properties. Evaluation of the efficacy of this treatment modality has been difficult because of variations in patients, dosages and treatment schedules, follow-up periods, and classes of interferon used. Interferon is administered as an intralesional injection to the dermal core of the lesion 3 times weekly for 3 weeks. Side effects have included fever, a flu-like syndrome, malaise, and myalgias. Women being treated with interferon should be instructed to avoid conception, as the long-term effects on the fetus or the woman's reproductive capacity are not known. The safety of this therapy has not been evaluated in patients less than 18 years of age.

Genital warts may resolve spontaneously. However, watchful waiting may be inappropriate because of the neoplastic potential of some subtypes of this virus.

SYPHILIS

Treponema pallidum, the causitive bacterium of syphilis, is spiral in shape and measures 6-20 by 0.1-0.18 μm, too small for visualization by light microscopy.

Clinical Manifestations

Approximately one-third of persons exposed through sexual contact to a partner with infectious syphilis will contract infection. Approximately 3 weeks after exposure, the primary lesion, known as the chancre, develops on the vulva, vagina, cervix or anus. The initial lesion is often single but multiple lesions may occur. The lesions are usually painless and associated with regional adenopathy. If they are left untreated, healing occurs in a few weeks. Hematogenous dissemination begins during this primary stage.

Secondary syphilis typically becomes manifest 4-10 weeks after the development of primary syphilis. The primary chancre is still present in 18-34% of patients presenting with secondary syphilis. Secondary syphilis may have a number of manifestations. Its most prominent and frequent feature is a rash, which occurs in 75-100% of infected patients and is nonpruritic, maculopapular and symmetric. It typically involves the trunk, extremities, palms and soles. Another cutaneous feature of secondary syphilis is condylomata lata, which is evident in 10-20% of patients and usually occurs in the anogenital region. Lesions may be broad-based, flat or heaped-up, have a fleshy, pearly gray appearance and are typically found near the site of the original chancre. Systemic symptoms often accompany the rash. These include headache, fever, musculoskeletal pains, gastrointestinal complaints, weight loss and low-grade fever. When no evidence of clinical disease exists, the patient is in the latent stage.

Tertiary syphilis occurs in one-third of patients, takes years to develop, and has not been reported in adolescents. Gummata, the chronic inflammatory lesions of tertiary syphilis, occur predominantly in skin, subcutaneous tissues or bone, although they have been described throughout the body and may occur at more than one site. Cardiovascular manifestations occur in about 10% of patients, the most common lesions being aortic aneurysm (40%), aortic regurgitation (29%) and coronary ostial stenosis (26%).

Involvement of the central nervous system occurs at all stages of syphilis. Clinical manifestations of neurosyphilis may include diffuse meningovascular, general paresis, tabes dorsalis and gumma of the brain.

Diagnosis

T. pallidum cannot be cultured. However, there are two methods for direct examination of the organism: dark field microscopy and fluorescent microscopy. Dark field microscopy is both sensitive and specific. It requires live organisms that can be recognized by their corkscrew morphology and flexing hairpin motility. To ensure adequate specimens, the lesions should be cleansed and scraped firmly to collect serum. The second method requires staining the specimen with fluorescein-labeled *T. pallidum* globulin and using a fluorescent microscope.

Serologic testing is the mainstay of syphilis screening and diagnosis. It is particularly useful when the patient does not have any clinical lesions that can be microscopically evaluated or when antibiotic therapy has reduced the sensitivity of dark field microscopy. There are two types of serologic assays, non-specific nontreponemal assays and specific treponemal assays. The two nontreponemal tests most widely used are the Venereal Disease Research Laboratory test (VDRL) and the rapid plasma reagin test (RPR). These tests are relatively inexpensive, easy to perform, and sensitive, making them useful for screening purposes and because they can be quantitated, they can be used to monitor the response to therapy.

Treponemal tests are used as confirmatory tests. Those most often used are the fluorescent treponemal antibody absorption test (FTA-ABS), the FTA-ABS double staining (DS) test, the microhemagglutination assay for antibodies to *T. pallidum* (MHA-TP), and the hemagglutination treponemal test for syphilis (HATTS). These tests are very sensitive and specific but are difficult to perform and relatively expensive in comparison with the VDRL and RPR. In most patients, the treponemal tests remain positive for life and do not necessarily indicate active *infection*.

Treatment

The CDC treatment guidelines for syphilis are outlined in table. Following treatment, patients should be evaluated monthly for the first 3 months, then 6, 9, 12 and 24 months later. The *nontreponemal* antibody titers should decrease >fourfold within 3 months. In patients with latent disease or in those who are reinfected, a decline in antibody titer may take up to 1 year. A fourfold increase in antibody titer demands retreatment.

Syphilis is a reportable disease. All sexual partners need to be notified and evaluated. Sexual partner(s) should be treated at the same time as the index case. For patients with primary syphilis, all partners within the preceding 3 months should be treated. For patients with secondary syphilis, all partners within the preceding 6-12 months are considered at high risk and should be treated.

GENITAL HERPES SIMPLEX VIRUS

Herpes simplex virus (HSV)-1 and HSV-2 are DNA viruses. HSV

infections are usually transmitted by intimate contact with an HSV-infected person who is shedding the virus from the oral or genital tract. Infections in the normal human host are usually limited to mucocutaneous sites and to the neuronal nucleus within the sensory *ganglion*. Initial infection occurs at the site of viral *inoculation* and subsequently spreads to contiguous cells and ensory and autonomic nerve *endings*. Replication in the dermis and epidermis occurs and the viral particles are transported to the *neuronal* nucleus where the virus becomes latent, persisting for life.

CLINICAL MANIFESTATIONS

Clinical *manifestations* of genital herpes depend on a variety of factors *including* the host response, previous infection with either HSV-1 or HSV-2, and viral type. The various clinical syndromes are described below, but it :s important to emphasize that recent seroepidemiologic studies have der..onstrated that clinically ev:dent genital HSV infections represent barely the tip of the iceberg. Whereas only about 5% of the population, clinical histories compatible with genital herpes, at least 20% of the general US population has serologic evidence of prior infection with HSV-2, the usual cause of genital disease.

A first episode is the initial clinical event and is classified as either a primary infection or a non-primary initial infection. Primary infections are those occurring in the total absence of HSV antibodies and nonprimary are those occurring in the presence of antibodies to HSV-1, HSV-2 or both. Clinical manifestations vary according to whether the infection is primary or nonprimary. The first episode of genital herpes may be accompanied by symptoms varying from mild to severe. In general, symptoms associated with primary infection are more severe, lasting an average of 3 weeks. Systemic symptoms commonly include headache, myalgias, fever and malaise; severe dysuria occurs more often in women than in men. Urinary retention may occur in women, and usually is secondary to pain or results from autonomic nerve involvement. Lesions usually will erupt externally and often look like small clusters of blisters filled with clear or whitish fluid on top of red inflamed skin. The vesicles progress through stages of ulcer formation, crusting and healing. Lesions may coalesce and form large ulcerations. Tender nonfluctuant inguinal lymphadenopathy usually appears in the second or third week of the illness and is frequently the last manifestation to resolve. HSV can be isolated from the cervix in 90% of primary infections in women and if untreated, cervical HSV may be cultured in some women for up to 3 months. Approximately 10% of patients have concomitant oral pharyngitis.

TABLE 7.4. CDC SYPHILIS TREATMENT GUIDELINES

Early syphilis

1. Preferred therapy: benzathine penicillin, 2.4 million units i.m. once
2. Alternative therapy: doxycycline, 100 mg b.i.d. or tetracycline, 500 mg q.i.d., either for 2 weeks orally
3. Therapy for patient who is allergic to both penicillin and tetracycline: skin-test and (if needed) desensitize or if compliance is assured:
   erythromycin, 500 mg q.i.d. orally for 2 weeks
   or ceftriaxone, 250 mg i.m. daily for 10 days
4. Therapy for patient who has clinical meningitis or neurologic findings: treat with regimen recommended for neurosyphilis

Late syphilis (greater than 1 year)

1. Preferred- therapy: benzathine penicillin, 2.4 million units i.m. weekly (3 doses)
2. Alternative therapy: doxycycline, 100 mg b.i.d. or tetracycline, 500 mg q.i.d., either for 4 weeks orally
3. Specific indications exist for lumbar puncture

Neurosyphilis

Preferred regimen: aqueous crystalline penicillin G, 12-24 million units daily (divided every 4 h) i.v. for 10-14 days, or

Aqueous procaine penicillin G, 2.4 million units i.m. daily, and 500 mg probenecid orally q.i.d., both for 10-14 days, either

May be followed by benzathine penicillin, 2.4 million units i.m. weekly (3 doses)

Syphilis in pregnancy

Treat with penicillin regimen appropriate for stage of syphilis; if penicillin-allergic, skintest and if needed desensitize

The duration of clinical illness in nonprimary infection is shorter than that of primary infection. The severity of the infection is influenced by the viral inoculum, the virulence of the infecting strain, and the status of the patient's immune response.

Recurrent episodes of genital herpes usually are milder than the first episodes. Lesions may be present anywhere in the genital track or on the buttocks, thighs or sometimes the back. A recurrent episode usually completely heals in 8-12 days. The average rate of recurrence is about four episodes per year for the first few years, commonly followed by a gradual decrease in frequency. At least 50% of patients report prodromal symptoms such as pain, itching, burning, numbness, tenderness, tingling, and sacral anesthesia or shooting pains and discomfort in the affected area. These may occur 2-3 days before eruption of the lesions. During the early stage of the disorder there may be a small, itchy red area that is painful or sensitive to touch. The lesions develop into papules and/or vesicles that contain clear or white fluid and often occur in groups. Healing begins with the formation of a crust-like area, which falls off after 5-10 days leaving smooth nontender skin. Stress, trauma, menses, influenza, chronic illness, and ultraviolet light have been found to trigger genital herpes.

HSV infection of the rectum is common in homosexual men and heterosexual women who engage in receptive anal intercourse. Symptoms include anorectal pain, rectal discharge, tenesmus and hematochezia. Most symptomatic proctitis is associated with primary HSV-2 infection.

Diagnosis

Viral culture is the most sensitive and specific diagnostic test for HSV infection. Viral isolation is possible from about 90% of vesicles and pustules in contrast to 30% from crusted lesions. Vesicles or pustules should be broken and their bases swabbed vigorously with a dacron or cotton-tipped swab. Assuming prompt transport to the laboratory followed by immediate tissue culture inoculation, HSV usually will grow within 1-4 days. Typing of the virus may be important and can be performed with commercially available monoclonal antibodies. Although the clinical disease is usually the same, the probability of recurrences is determined by the type of infecting virus. Specifically, HSV-1 genital infections rarely recur whereas HSV-2 infections commonly recur.

Alternatives to viral culture include cytologic diagnosis by Papanicolaou or Tzanck stains and viral antigen or DNA detection. The antigen detection techniques are less expensive and more rapid than viral culture, but their sensitivity in the best of hands rarely exceeds 90%. Cytologic diagnosis is much less sensitive than viral culture and cannot differentiate among HSV-1, HSV-2, and varicella zoster virus (VZV).

Treatment

The antiviral acyclovir (ACV) is the mainstay of therapy for genital HSV infection. The treatment guidelines for genital HSV are outlined in table. Treatment of first-episode genital HSV with either intravenous, oral or topical acyclovir has been shown to shorten the clinical course and the duration of lesions significantly and reduce the duration of viral shedding. Oral and intravenous ACV apear to have greater antiviral efficacy, and unlike topical therapy, oral and intravenous therapy prevent new lesion formation and shorten the duration of systemic symptoms and viral shedding. Oral or intravenous acyclovir is not recommended for use during pregnancy except in life-threatening HSV or VZV infections. In the US, there are no other antivirals available for therapy of genital HSV. Compounds available elsewhere include vidarabine monophosphate, isoprinosine and foscarnate.

CONCLUSIONS

Sexually transmitted diseases have reached epidemic proportions in sexually active adolescents who are initiating sexual intercourse at younger ages, often with multiple partners and concomitant illicit substance use. The high rates of STDs in adolescents are accompanied by high rates of serious medical complications, including PID, infertility, ectopic pregnancy, cervical carcinoma, and congenital infections in their newborns.

The overriding goal of STD intervention programs is to prevent the spread of these organisms. This is particularly difficult in adolescents, who may have a strong sense of invulnerability and little concern for long-range consequences of their behavior. Strategies for preventing STDs should include informing adolescents about the risks of all forms of sexual intimacy while providing them with the knowledge and methods for prevention of infection. Early intervention to modify adolescents' perceptions and interactions with peers has been proposed as an effective way to prevent the onset of alcohol and drug abuse as well as of sexual activity and non-use of condoms.

Teenagers should be instructed about safe sex and be encouraged to use 'barrier contraceptives with every coital act. Sexually active teenagers should be screened yearly, and high-risk adolescents such as runaways, drug abusers and those with multiple sexual partners more frequently.

More clinical services for diagnosing and treating STD are needed and must be readily accessible. Mobile clinics or vans can provide services to the growing number of homeless youth. STD services can also be integrated into programs for family planning, prenatal care, and AIDS prevention to ensure that adolescents seeking such health care also receive routine screening and treatment for STD. Expanded clinical and health education services should be available to youth in and out of schools.

8

ANTIBIOTIC THERAPY

New antibiotics are constantly being developed, investigated, and marketed. The successful antibiotics are those that find a particular niche, by expanding the spectrum of activity beyond existing agents, by compensating for bacterial resistance to older agents, or by possessing pharmacodynamic properties which give them advantages over available preparations. In this review, we will discuss some new agents which have become available over the past few years, new ways of using some old antibiotics, and selected antibiotics on the horizon which appear promising.

BETA-LACTAMASE INHIBITORS

From the outset of the antibiotic era, microorganisms have developed means to avoid antibiotic-mediated killing almost as fast as new antimicrobials have been developed. Microorganisms have multiple mechanisms of resistance, although production of enzymes (β-lactamases) that hydrolyze the β-lactam ring is the most important and best understood mechanism of bacterial resistance. For example, when penicillin was brought into widespread use in the early 1940s, staphylococci, as well as streptococci, were uniformly susceptible. Shortly thereafter, staphylococci were identified which produced a penicillinase that inactivated the drug, rendering the organism resistant to the antibiotic's effects. Within a few years, most hospital isolates of staphylococci were resistant to penicillin.

One approach to counteract β-lactamase production has been the development of antibiotics that are inherently more resistant to degradation by these enzymes. Oxacillin and nafcillin, for example, are relatively resistant to the actions of penicillinase and thus are effective antibiotics against many *Staphylococcus aureus*. (Of course, the microbe always staying one step ahead, many *S. aureus* are now resistant to these antibiotics by mechanisms not involving Q-lactamase production.) Similarly, the activity of cefuroxime and the third generation cephalosporins against β-lactamase producing strains of *Haemophilus infuenzae* is due to the inherent ability of these antibiotics to resist enzymatic destruction.

An alternate approach has been the development of β-lactamase inhibitors. Three such antibiotics are in widespread clinical use and several more are under investigation. The available drugs are amoxicillin-clavulanate, ticarcillin-clavulanate, and ampicillin-sulbactam. Clavulanate and sulbactam are themselves β-lactam antibiotics, albeit with weak antimicrobial activity. They are, however, avidly bound by β-lactamases, acting as 'suicide inhibitors'. This protects the combined β-lactam antibiotic from being enzymatically degraded and thus expands its spectrum of activity.

In general, the β-lactamase inhibitor compounds are more active than the parent drug against β-lactamase-producing organisms such as *S. aureus, H. influenzae, Moraxella* (formerly *Branhamella) catarrhalis, Neisseria gonorrhoeae,* and *anaerobes*.

However, the β-lactamase inhibitors are not a panacea, because all β-lactamases are not alike. These enzymes are produced by different microorganisms and have varying affinities for different antibiotics. Most notably, the cephalosporinases produced by some strains of *Pseudomonas aeruginosa, Serratia marcescens,* and *Enterobacter cloacae* are generally not inhibited by sulbactam or clavulanate. In addition, some enterobacteriaceae are able to produce β-lactamase in high enough quantities to overcome the effects of the P-lactamase inhibitor.

Clinically, amoxicillin-clavulanate is a particularly useful oral pediatric antibiotic. As a result of its expanded coverage against β-lactamase-producing strains of *H. influenzae* and *M. catarrhalis,* it is an excellent second-line agent for children with acute otitis media or sinusitis who fail therapy with amoxicillin. It is the drug of choice for the therapy (or prophylaxis) of wound infections due to animal or human bites because of its activity against staphylococci, streptococci, oral anaerobes; *Eikenella corrodens,* and *Pasteurella multocida*. It has also proven efficacious in treating lower respiratory tract infections, skin and soft tissue infections, and urinary tract infections, though a more narrow spectrum antibiotic is often appropriate. The drug is safe, although about 10% of 'recipients experience gastrointestinal side effects, particularly diarrhea.

Ampicillin-sulbactam is the parenteral counterpart of amoxicillin-clavulanate. Its spectrum of activity suggests it may be useful if parenteral therapy is indicated for the therapy of head and neck, respiratory tract, skin and soft tissue, or bone and joint infections. However, few comparative clinical trials have evaluated its use, and except for serious bite wound infections, alternate effective therapy with a single agent is available for these infections. There are major concerns about sulbactam's (and clavulanate's) ability to achieve adequate cerebrospinal fluid concentrations. Therefore, this antibiotic should not be used in infants or children with suspected sepsis/meningitis.

Ticarcillin-clavulanate has been used most commonly in pediatrics in combination with an aminoglycoside as part of empiric antibiotic coverage for the febrile, neutropenic cancer patient, although it is not as efficacious as vancomycin for gram-positive infections. Clinical trials have also established its efficacy in treating *S. aureus* skin, soft tissue, bone, and joint infections, although a narrower spectrum antibiotic would usually be more appropriate. To reemphasize, while ticarcillin-clavulanate broadens ticarcillin's gram-positive spectrum and increases its coverage against anaerobes, *H. influenzae,* and *Klebsiella pneumoniae,* it does not reliably improve ticarcillin's coverage against many enterobacteriaceae, including pseudomonas. Therefore, like ticarcillin, it should generally be used in combination with an aminoglycoside for serious gram-negative infections.

ORAL CEPHALOSPORINS

The oral cephalosporins are widely used in pediatrics. The first-generation oral cephalosporins, such as cephalexin and cephadroxil, have excellent activity against gram-positive bacteria and thus have earned a niche in the therapy of skin and soft tissue infections and bone and joint infections. Cefaclor, a second-generation cephalosporin, is commonly used for the therapy of acute otitis media and respiratory infections although its activity against β-lactamase-producing strains of *H. influenzae* and *M. catarrhalis is* unreliable. In addition, its use is associated with an approximately 2% incidence of serum sickness. Five new oral agents have recently become available to expand our ability to manage selected infections in ambulatory pediatrics.

Cefuroxime axetil is an oral preparation of the widely used intravenous drug. It is β-lactamase stable and thus would appear to be suitable therapy for acute otitis media and sinusitis. It also is more active against *S. aureus* than is cefaclor, and can be dosed on a twice-daily basis. However, no oral suspension of cefuroxime axetil has yet been marketed because of its bitter taste, and therefore it cannot be used in young children who are unable to swallow pills.

Cefprozil and cefpodoxime, new oral cephalosporins, and loracarbef, the first available carbacephem antibiotic, all have in vitro activity against the common pathogens causing otitis media and sinusitis. Of the three, cefpodoxime appears to be the most reliably active against β-lactamase-producing strains. Each appears to be clinically effective against otitis media, although further data, and more clinical experience, are necessary to determine each of the drug's specific benefits and shortcomings, and their place in pediatric therapeutics. All three are dosed on a twice-daily basis.

Cefixime is the first oral third-generation cephalosporin. It has superior activity against β-lactamase-producing organisms and thus is excellent therapy for cases of otitis media which have not responded to amoxicillin therapy. However, it probably should not be employed as primary therapy because it appears to have decreased efficacy against pneumococci as compared to amoxicillin. Cefixime has no useful activity against staphylococci and so should not be used for skin or soft tissue infections where staphylococci are likely pathogens. Cefixime does have expanded coverage against many gram-negative organisms (though not pseudomonas) and therefore may be useful for therapy of complicated urinary tract infections (UTI), or in patients with UTIs who have a sulfa-allergy. As a result of a long halflife, cefixime may be dosed on an every 24-hour basis.

INTRAVENOUS CEPHALOSPORINS

The intravenous cephalosporins have earned a major role in pediatric antibiotic therapy. While space does not permit a comprehensive review of these antibiotics, the interested reader can refer to an excellent recent review. The first-generation cephalosporins, e.g. cefazolin have earned a role in the therapy of skin and soft tissue infections, bone and joint infections, and surgical prophylaxis because of their excellent gram-positive coverage. Cefuroxime, a second-generation cephalosporin, has earned a place in the therapy of common pediatric infections such as pneumonia, pyarthrosis, and periorbital or buccal cellulitis as a result of its activity against the common pathogens: staphylococci, streptococci, and β-lactamase-producing strains of *H. influenzae*. The third-generation cephalosporins sacrifice some grampositive activity for expanded gram-negative activity. They penetrate the cerebrospinal fluid well, and cefotaxime and ceftriaxone have become mainstays in the empiric therapy of bacterial sepsis/ meningitis. Ceftriaxone has secured a role in the out-patient therapy of selected serious pediatric infections because its prolonged elimination half life allows once-daily dosing. Ceftazidime has earned a niche among the third-generation drugs because of its anti-pseudomonal activity and has proven efficacious as therapy in febrile neutropenic cancer patients, in cystic fibrosis patients, and in hospitalized patients with nosocomial infections.

Two new cephalosporins, cefpirome (HR 810) and cefepime (BMY 28142), can be considered the first fourth-generation cephalosporins. They share with the third-generation drugs a high degree of activity against gramnegative organisms and β-lactamase stability, and like ceftazidime have high activity against ***P. aeruginosa***. However, they also retain gram-positive coverage, with good activity against methicillin-susceptible staphylococci and nonenterococcal streptococci. They do not have activity against *Bacteroides fragilis*. In an open trial, cefepime proved efficacious against a variety of gram-positive and gram-negative infections, primarily of the respiratory tract and skin and soft tissue. Comparative trials need to be performed to establish their possible role in infections in granulocytopenic cancer patients and cystic fibrosis patients, and in the therapy of nosocomial infections.

CARBAPENEMS

Imipenem was the first carbapenem antibiotic and is available combined with cilastatin. Cilastatin inhibits a renal brush-border enzyme that inactivates imipenem, thereby increasing urinary excretion and decreasing nephrotoxicity. Imipenem/cilastatin has the broadest spectrum of activity of any available antibiotic with good gram-positive coverage, broad gramnegative coverage including pseudomonas, and excellent anaerobic activity including *B. fragilis*. Indeed, it is easier to list the bacteria that imipenem does not cover - namely *Enterococcus faecium, Xanthomonas maltophilia, Clostridium difficile, mycoplasma, chlamydia*, and many methicillin-resistant strains of *S. aureus* and coagulase-negative staphylococci.

Clinically, imipenem has proven efficacious in treating a wide variety of pediatric infection. However, it generally should not be used to treat community acquired infections and instead should be reserved for therapy of nosocomial infections with organisms resistant to other antibiotics, and for polymicrobial infections such as intraabdominal sepsis. An infectious diseases consultation should be obtained in most such cases. As a result of its broad spectrum of activity, imipenem has also been investigated as empiric monotherapy in febrile neutropenic cancer patients, although sufficient data are not yet available to recommend it for this use.

Problems with its use include disruption of the patient's normal microbial flora including anaerobes. This can lead to superinfection with fungi or resistant bacteria. Resistance to imipenem has developed during therapy of pseudomonas infections, so the concomitant use of an aminoglycoside may be desirable. The major adverse effect noted during therapy in adults has been a 1-3% incidence of seizures. A high incidence of seizures was also reported in imipenem-treated children with meningitis.

MONOBACTAMS

Unlike other β-lactam antibiotics, the monobactam antibiotics are monocyclic, without a second ring attached to the β-lactam ring. Aztreonam is the first monobactam antibiotic to be available clinically.

Azireonam has a narrow spectrum of activity, limited to aerobic gramnegative bacilli, including most strains of pseudomonas, the common gastrointestinal pathogens, and the common pediatric pathogens *Neisseria meningitides* and *H. influenzae*. It is stable against β-lactamase-producing organisms. Since aztreonam has neither gram-positive nor anaerobic coverage, it must be used either as directed therapy against known pathogens or in combination with another antibiotic.

Clinical trials have established the efficacy of aztreonam for a number of different pediatric infections. It has been studied most frequently in urinary tract infections, where it has resulted in excellent clinical and microbiologic outcomes. A variety of other serious gram-negative infections have been treated with aztreonam including skeletal infections, peritonitis, and pulmonary exacerbations in cystic fibrosis patients. Responses of children with bacteremia and meningitis (the majority due to *H. influenzae* type b) have been satisfactory. Aztreonam (in combination with ampicillin) was effective and safe when compared with ampicillin and amikacin for empiric therapy of neonatal sepsis.

Most of our current information on aztreonam has resulted from open, noncomparative trials. Comparative studies in children must be performed to determine more precisely its role in pediatric therapeutics. Clearly, however, aztreonam may have a niche in certain clinical situations. While its spectrum of activity and clinical efficacy are similar to that of the aminoglycosides, aztreonam lacks the aminoglycosides' oto- or nephrotoxicity. Therefore, it is an attractive agent for gram-negative infections in patients with underlying renal insufficiency or who are on other nephrotoxic drugs. In addition, because of its structure, aztreonam shares minimal immunologic cross-reactivity with other β-lactam agents and may be used safely in penicillin-allergic patients.

MACROLIDES

Over the years, erythromycin has proven to be a valuable antibiotic against gram-positive bacteria as well as against common respiratory pathogens such as pertussis, chlasnydia, legionella, and mycoplasma. A new series of macrolides have been developed and investigated around the world over the past few years. These agents are notable not so much for any significant differences in their spectrum of activity, but more for pharmacodynamic properties which make them potentially useful additions to our antibiotic armamentarium. We focus here on two which appear to be particularly promising: azithromycin and clarithromycin.

Azithromycin is somewhat less active in vitro than erythromycin against most streptococci and staphylococci. However, it is several-fold more active against a wide range of upper and lower respiratory tract pathogens such as *H. influenzae, M. catarrhalis, Legionella pneumophilia,* and *Mycoplasma pneumoniae*. Clarithromycin has somewhat higher activity than does erythromycin against susceptible strains of streptococci and staphylococci, legionella, and *M. catarrhalis*. It is slightly less active against haemophilus species.

Again, it is the pharmacokinetics of the new macrolides which make them such promising antibiotics. After oral administration of azithromycin, only about 40% of the compound is bioavailable. Peak serum concentrations are exceptionally low, but intracellular drug concentrations are 10- to 100-fold higher. Clarithromycin is about 55% bioavailable after oral administration. Like azithromycin, it diffuses very well into body tissues, achieving tissue concentrations that are 10- to 20-fold higher than serum concentrations. A half-life of 2.5-6 h allows clarithromycin to be dosed twice daily, while the metabolism and elimination of azithromycin are very complex with an extended tissue half-life on the order of 2-3 days. This allows azithromycin to be dosed once daily, and suggests that a 5-c course may be sufficient for skin and soft tissue or respiratory tract inf tions, since adequate tissue levels are maintained for an additional 4-7 d, after the drug is discontinued.

As a result of their spectrum of activity and high tissue concentratioi the new macrolides are expected to be useful agents for a variety of infi tions. A single 1-gram dose of azithromycin appears to be effective theca, for uncomplicated genital chlamydial infections. Studies in adults with i fections of the respiratory tract or skin and soft tissue have been promisir and these agents are now being investigated for pediatric infections includii otitis media.

In addition, the new macrolides have activity against some other path gens which have emerged as clinical problems in immunocom-promise hosts. As a result of their ability to achieve high concentrations in cells, the are being investigated in HIV-infected patients as therapy against intracelli lar pathogens such as *Mycobacterium avium* complex and toxoplasmosis.

AMINOGLYCOSIDES

The aminoglycoside antibiotics (gentamicin, tobramycin, netilmici: amikacin, etc.) have fallen somewhat into disfavor, largely because concerns of nephro- and ototoxicity. The third-generation cephalosporin with their broad spectrum of activity against gram-negative organisms ar their relative safety, have rightly assumed a growing role in the therapy proven or suspected gram-negative infections. Nonetheless, aminoglycosid have advantages which make them valuable antibiotics: (1) they are high active, with a dose-dependent response; (2) the development of resistan during therapy is rare; (3) synergistic killing may occur with concomita f3-lactam antibiotic use, and (4) they are relatively inexpensive.

Research over the past decade has changed our thinking about optim dosing of aminoglycosides. Gilbert recently summarized the concep behind less frequent dosing, particularly once-daily dosing using the san total daily dose (e.g. giving children 7.5 mg/kg q 24 h rather than 2.5 mg/kg 8 h). Theoretically, the advantages of such a regimen include: (1) high efficacy due to achievement of higher serum levels (both bacterial killing at clinical efficacy against gram-negative infections correlate with higher pe, blood levels), and (2) less toxicity (in animal models, nephrotoxicity at ototoxicity seem to correlate with the absence of drug in the target organ during part of the dosing interval, i.e. low trough levels). When the drug is administered once daily, serum levels are undetectable for up to 8 h. This does not seem to adversely affect efficacy due to a postantibiotic effect which suppresses bacterial growth even after serum levels fall below the minimum inhibitory concentration. The duration of the postantibiotic effect correlates with peak serum levels. An additional theoretical argument for less-frequent dosing is the concept of 'adaptive resistance'. In vitro and in animal models, a lag period occurs after the initial phase of bacterial killing, during which surviving bacteria become relatively impermeable to the aminoglycosides.

Human data using once-daily dosing of the aminoglycosides are sparse, with little data available in children. In the studies that have been published, efficacy appears to be equivalent to former regimens without increased adverse effects. While a change to every 24-hour dosing cannot be recommended without further studies in children, the evidence suggests that less frequent dosing (every 12 h) may be efficacious, less toxic, and cheaper.

QUINOLONES

The fluoroquinolone antibiotics are an exciting addition to the antibiotic world. They have a unique mechanism of action, inhibiting the bacterial enzyme DNA gyrase which interferes with the supercoiling of bacterial DNA. These drugs are notable for their marked bactericidal activity, their broad spectrum of activity, and their favorable pharmacodynamics.

Nalidixic acid was the first available quinolone, but low serum concentrations limited its use to urinary tract infections. Ciprofloxacin, the first available fluoroquinolone, has enabled us to see the potential utility of this class of antibiotics. It is well absorbed after oral administration (an intravenous form is also available), achieves high levels in the serum and in tissues, and has a broad spectrum of activity. Its gram-positive coverage is reasonably good though it does not have activity against enterococci and its staphylococcal coverage is better than its streptococcal coverage. Serious pneumococcal infections have been reported in patients on ciprofloxacin therapy, so it probably should not be used routinely for respiratory tract infections where the pneumococcus is a likely pathogen. In addition, while its coverage against staphylococci is good, including activity against some methicillin-resistant strains, increasing reports of the development of ciprofloxacin-resistance suggest that its utility for these infections may diminish as use increases. Ciprofloxacin has excellent gram-negative coverage and the quinolones are the only available oral agents active against *P. aeruginosa* Ciprofloxacin does not have significant anaerobic coverage.

Ciprofloxacin may be useful in the therapy of a number of clinical syndromes. Not surprisingly, since its main route of excretion is the kidneys, many clinical trials in adults have demonstrated its efficacy in treating urinary tract infections. It has been used successfully for bone and joint infections (particularly in the diabetic foot) because of its activity against staphylococci and gram-negative organisms. Ciprofloxacin may also play a role in the therapy of bacterial gastroenteritis, as the quinolones are the only antibiotics available that have activity against all the major pathogens (salmonella, shigella, campylobacter, yersinia, enterotoxigenic *Escherichia coli*).

As a result of its high penetration into tissue, ciprofloxacin has been used to eradicate the chronic carrier state of *Salmonella typhi*. Similarly, one study showed that a single dose of ciprofloxacin successfully eradicated carriage of meningococcus. Ciprofloxacin also has activity against mycobacteria, and with its intracellular activity has been included in some multiple-drug regimens to treat *M. avium* infections.

In addition, ciprofloxacin is active against penicillinase-producing strains of *N. gonorrhoeae*. A single oral dose of 500 mg ciprofloxacin has proven highly efficacious in treating uncomplicated gonococcal urethritis and cervicitis, though this is not sufficient therapy for *Chlamydia trachomatis*.

Although unreliable activity against pneumococci mediates against its routine use for the therapy of respiratory tract infections, ciprofloxacin has been used successfully to treat pulmonary exacerbations in cystic fibrosis patients in whom pseudomonas, staphylococcus, and haemophilus are the predominant pathogens.

As a result of its oral bioavailability, activity against a wide range of gram-negative organisms including pseudomonas, and lack of anaerobic activity, ciprofloxacin has been investigated for prophylaxis of granulocytopenic cancer patients. Several studies have demonstrated a decrease in gramnegative infections in patients with prolonged neutropenia receiving prophylactic ciprofloxacin, but there are insufficient data to make specific recommendations at this time.

Unfortunately, ciprofloxacin is not approved for pediatric use because of cartilage damage observed with quinolone use in juvenile animals. It is not clear if this effect is seen in humans, and there are no reports of cartilage damage associated with quinolone use in children. A recent study in young cystic fibrosis patients on long-term ciprofloxacin, for example, revealed no evidence of cartilage damage by magnetic resonance imaging. Nonetheless, until larger, careful studies are performed, ciprofloxacin is not recommended for use in children under 18 years of age.

A number of other fluoroquinolone antibiotics have been investigated. Several, including norfloxacin, ofloxacin and lomefloxacine have been licensed for use in the United States. Norfloxacin is less active than ciprofloxacin against *P. aeruginosa* and *S. aureus*. Like ciprofloxacin, it has proven highly efficacious in the therapy of gastrointestinal and urinary tract infections. Ofloxacin has in vitro activity comparable to that of ciprofloxacin and greater than that of norfloxacin. Its anti-mycobacterial activity has earned it a role in the therapy of drug-resistant tuberculosis. Ofloxacin is somewhat less active than ciprofloxacin against enterobacteriaceae including pseudomonas. Overall, its safety profile and clinical efficacy have been comparable to ciprofloxacin in comparative trials. Another new quinolone, temafloxacin, was withdrawn from the market shortly after licensure due to the occurrence of serious adverse reactions (hemolysis, thrombocytopenia, renal failure, and several deaths) which had not been apparent in pre-licensure evaluation.

GLYCOPEPTIDES

Teicoplanin is a glycopeptide antibiotic that resembles vancomycin in structure, mechanism of action, and spectrum of activity. It is active against gram-positive organisms such as streptococci, enterococci, and staphylococci (including coagulase-negative and methicillin-resistant strains). Its similarities in structure and mechanism of action indicate that teicoplanin generally would not be expected to have activity against gram-positive organisms that are vancomycin resistant.

Teicoplanin holds several potential advantages over vancomycin. First, the incidence of nephrotoxicity appears to be reduced with teicoplanin compared to vancomycin. Second, a longer half-life may permit once-daily dosing. Finally, it can be administered intramuscularly. These last two properties make teicoplanin a promising drug for out-patient therapy without the need for prolonged intravenous access.

Teicoplanin's spectrum of activity suggests that it may be useful in the therapy of skin and soft tissue infections, bone and joint infections, and catheter-related gram-positive infections. Teicoplanin has been evaluated in multiple clinical trials in adults. Early studies were somewhat disappointing, probably due to the fact that very low doses (less than 6 mg/kg) were used. Subsequent studies using larger doses have shown good clinical efficacy in treating a variety of gram-positive infections, such as bone and joint, skin and soft tissue, or respiratory tract infections.

Very little data are available concerning the use of teicoplanin in children. Approximately 200 children have participated in open, noncomparative trials in Europe. The drug proved safe, and clinical efficacy was good in these patients (neutropenic and not neutropenic) with a variety of infectious syndromes.

CONCLUSION

As new antibiotics are introduced into the market with flashy advertising and bold claims of superiority, the clinician may feel overwhelmed. The pediatrician or practitioner caring for children may have even more difficulty as oftentimes less is known about the antibiotic's effects in children than in adults. Even after licensure, much remains to be learned about a new antibiotic including the possibility of rare adverse effects, and previously unrecognized limitations on its activity and efficacy. Often, new antibiotics do not hold distinct advantages over existing agents. Therefore, while it may be tempting to use these antibiotics as soon as they become available, the prudent physician may opt to wait for further data from comparative clinical trials in pediatric infections, and for experience with the new agent to grow.

9

Treatment of Parasitic Diseases

Present levels of travel, changing immigration patterns and the spread of HIV infection have given renewed prominence to parasites even in developed countries. In this review, doses of antiparasitic drugs are mostly culled from the Medical Letter, an invaluable concise reference which is updated every 2 years. Informed advice on parasitic disease is available through the Parasitic Disease Drug Service of the US Centers for Disease Control (CDC), Altanta, Ga., USA.

PROTOZOA

Amebiasis (Entamoeba histolytica)

Different drugs act at different sites of amebic replication, namely, bowel lumen (diloxanide furoate, paromomycin, diiodohydroxyquin; (lodoquinol), intestinal submucosa (tetracycline, erythromycin), extraintestinal sites such as liver (chloroquine) and brain. Some drugs (metronidazole, tinidazole, ornidazole, emetine hydrochloride, 2-dehydroemetine) work on all tissues. A lumenal amebicide is always needed regardless of the disease site. No regimen is always effective in eradicating intestinal infection hence follow-up stool examination is necessary.

Asymptomatic Cyst Passers

The risk of asymptomatic cyst passers becoming symptomatic is small. Until recently the accepted view was that asymptomatic carriers are infected with strains of selected stable zymodemes (electrophoretic motility profiles) thought not to cause disease. There is now in vitro evidence that nonpathogenic zymodemes can become pathogenic during coculture with irradiated viable bacteria, although whether this can happen in vivo is not known. In nonendemic areas all cyst carriers should be treated.

TABLE 9.1. DRUG DOSES IN AMEBIASIS

Asymptomatic cyst passers

Diloxanide furoate 7-10 mg/kg/8 h (max 500 mg) for 10 days, or Diiodohydroxyquin (lodoquinol) 10 mg/kg/8 h (max 650 mg) for 20 days, or Paramomycin 8-10 mg/kg/8 h (max 500 mg) for 7 days

Invasive colitis

Metronidazole 15 mg/kg/8 h p.o. or i.v. for 7-10 days, or Tinidazole 50 mg/kg p.o., once daily for 3 days, *or* Ornidazole 40 mg/kg p.o., once daily for 3 days *Plus a* lumenal agent

Severe disease

Treat as for invasive colitis: consider adding dehydroemetine HCl 1-1.5 mg/kg/day (max 90 mg) i.m. for 2-3 days or emetine 1 mg/kg (max 60 mg) i.m. for 3 days

Absolute Indications for Treating Asymptomatic Cyst Passers

Amebiasis should be excluded in inflammatory bowel disease especially if considering steroid therapy which may induce severe amebic colitis, or if considering surgery. Other groups include HIV positive individuals in whom the adherence lectin of *E. histolytica* may stimulate viral replication, and the very young or malnourished.

INVASIVE INTESTINAL DISEASE

Extraintestinal Disease

Liver disease

Liver disease metronidazole is used. Chloroquine may be added because of the high concentrations reached in the liver. It has little activity elsewhere. Needle aspiration of the liver under ultrasound control is indicated for illness so severe as to require intervention pending a response to drugs, impending rupture, or lack of response within"5 days of commencing treatment. Ultrasound follow-up is useful.

Bowel perforation/peritonitis

Bowel perforation/peritonitis affected bowel should not be resected. Ensure adequate peritoneal drainage, scrupulous fluid and electrolyte balance, and rely on antiprotozoal and antibacterial therapy.

Free-Living Amoebae

Naegleria fowleri causes primary amebic meningoencephalitis (PAM) usually related to fresh water swimming. Survival is rare. Treatment is with i.v. amphotericin B 1 mg/kg/day for an unspecified period. Creditable alternatives include intrathecal and high-dose systemic miconazole, or systemic sulfizoxazole with rifampicin. Immunotherapy with hyperimmune serum or a monoclonal antibody may prove to be useful in the future.

Acanthamoebae cause granulomatous amebic encephalitis (GAE) and keratitis associated with soft contact lens use. The best treatment for GAE is unknown although pentamidine is used. Other drugs to consider include ketoconazole, miconazole, paramomycin, neomycin, 5-FC, and amphotericin B. Keratitis requires aggressive debridement plus topical 0. A propamidine isethionate every 30-60 min for 3 days, then × 10 daily for 4 weeks, plus topical 1% miconazole and neosporin. Oral itraconazole may also be helpful.

MALARIA

Introduction

Plasmodium vivax, Plasmodium ovale, and *Plasmodium malariae* infections are rarely fatal but nonetheless cause much morbidity. They are nearly all chloroquine sensitive, although resistence exists in isolated areas. *Plasmodium falciparum* remains the most important parasitic infection of the tropics. It causes potentially fatal cerebral malaria, and is ingenious in developing drug resistance. Chloroquine resistance is found in all malarious areas except for central America north of Panama, Mexico, Haiti, Dominican Republic, Egypt and the Middle East.

Malaria Prevention

Chemoprophylaxis cannot guarantee complete protection against malaria,· hence personal protection against mosquito bites as an adjunct to chemotherapy has assumed prominence. Such measures include sleeping in properly screened, air-conditioned rooms; mosquito nets (± impregnated with permethrin) around the bed; using mosquito coils or electric mat vaporization of synthetic pyrethroids overnight; wearing long-sleeved shirts and long trousers after dusk; using personal insect repellant, preferably one containing diethyl toluamide (= DEET, a compound that rarely causes toxic encephalopathy) or dimethyl phthalate; avoid perfumes and colognes. Short-term visitors to low-risk areas such as large towns may not need an, chemoprophylaxis if these measures are followed.

TABLE 9.2. SUGGESTED MALARIA CHEMOPROPHYLAXIS FOR TRAVELERS

No evidence of chloroquine or DHFR inhibitor resistance (Mediterranean. Middle Ea! parts of Central and South American, Caribbean) Chloroquine base - 5 mg/kg/week (max 300 mg), or Proguanil -3 mg/kg/day (max 200 mg)

Widespread chloroquine resistance (SE Asia, Pacific, Papua New Guinea, Africa) Mefloquine (250 mg) - 5 mg/kg/week; 15-19 kg 1/4 tablet/week; 20-30 kg 1/2 table week; 31-45 kg 3/4 tablet/week; > 45 kg I tablet/week, *or* Doxycycline - >. 8 years old 2 mg/kg/day (max 100 mg)

Standby treatment in areas of chloroquine resistance
Fansidar- 2-11 months 1/4 tablet; 1-3 years 1/2 tablet; 4-8 years I tablet; 9-14 years tablets, > 14 years 3 tablet; given once

Chemoprophylaxis

Commence prophylaxis 1-2 weeks before travel to achieve steady stat concentrations and to test for idiosyncratic reactions. Continue for 4 week after leaving the area. Chloroquine and mefloquine are usually prescribed a weight of base while quinine and quinidine are prescribed as weight of sail

Chloroquine: the drug of choice for sensitive *P. falciparum* and the othe human malarias, although regrettably in some areas (e.g. Papua New Guinea PNG) resistance has recently been demonstrated to *P. vtvax*. Toxicity Occasional - pruritus (especially in blacks) responds to steroids but no antihistamines, gastrointestinal side effects, headache, confusion, alopecia exfoliative dermatoses. Rare - neuropathy, myopathy. Retinopathy is exceedingly rare in children, probably requiring a cumulative dose of > 100 g.

Proguanil: a biguanide, which acts through its metabolite cycloguanil by selectively inhibiting malarial dihydrofolate reductase (DHFR). It is the safest of the antimalarial drugs. Toxicity: Occasional - mouth ulcers, abdominal discomfort and vomiting, alopecia, asymptomatic mild ineutropenia.

Pyrmethamine/dapsone (Maloprim): combines 12.5 mg pvrimethamine (P) with 100 mg of dapsone (D), acting by sequential inhibition of dihydropteroate synthetase (dapsone), and DHFR (pyrimethamine). Toxicity: With twice weekly use (t„2 of D is 22 h) agranulocytosis occurred in 1:2,000 to 1:5,000 users. It is now used once weekly, agranulocytosis occurring in 1:20,000 users of which 1/3 die. Only recommended in SE Asia and PNG although mefloquine or doxycycline are rapidly replacing it. Rare side effects are methemoglobinemia, megaloblastic anemia, leukopenia.

Mefloquine: a quinoline-methanol compound structurally resembling quinine. To forestall drug resistance in some countries it has been combined with Fansidar, a drug with serious side effects and a much shorter half-life (t_{12}). The combination is not recommended. Mefloquine should be used alone, probably as the drug of choice in areas with multiresistant *P. falciparum* although sporadic resistance has already been recognised. Terminal t„$_2$ in healthy adults is 1-4 weeks (shorter in those with malaria). Until 1991 it was administered as a weekly dose for 4 weeks, then every alternate week. Prophylaxis failures in the second week of this regimen have led the CDC to advise ongoing weekly doses. It is not recommended for children < 15 kg because of vomiting and erratic absorption, those with epilepsy or a psychiatric disorder. Toxicity: Occasional - nausea, vomiting, diarrhea, vertigo, confusion. Psychosis or convulsions are very rare.

Doxycycline: the tetracylines were recognized in 1952 as antimalarials, however it was only in the 1970s that they came into popular use. Toxicity: GI disturbances, exaggerated sunburn (use a sunscreen), candida vaginitis. Contraindicated in children < 8 years. 'Peptobismol' and antacids decrease bioavailability.

Amodiaquine: in 1988 the WHO advised against its use because of agranulocytosis and hepatotoxicity.

Fansidar (25 mg of P with 500 mg sulfadoxine): it is no longer recommended for prophylaxis because toxic epidermal necrolysis, Stevens-Johnson syndrome or erythema multiforme occur in about 1:8,000 users with mortality of 1:11,000-1:25,000 users. It is still occasionally recommended for treatment.

PROPHYLAXIS FOR THE BREAST-FED CHILD

Ineffective and harmless amounts of antimalarials are transferred into breast milk. Infants should receive prophylaxis according to weight.

Vaccination

The fact that effective partial immunity occurs in young adults in endemic areas plus the success of vaccines in laboratory animals suggested that a successful human vaccine was imminent. Regrettably field trials of anti-*P. falciparum* vaccines have been disappointing although the search goes on. An intriguing possibility in the future is immunisation against the mediators of serious disease (probably cytokines), particularly tumour necrosis factor (TNF), rather than the parasite. If the molecules inducing TNF production can be identified, antibodies against them may form the basis of an 'antitoxic' vaccine.

Treatment of Malaria

In general, a well child can be treated by mouth, whereas sicker children require parenteral therapy. Infection of a nonimmune host with *P. falciparum* may threaten life within hours. In travelers, assume suspected malaria to be chloroquine resistant, unless the child has not been in areas with reported resistance.

Chloroquine: see 'Chemoprophylaxis' above. Hypotension and heart block occur with high blood levels. Continuous infusion or srr__ it doses at short intervals avoid these problems.

Primaquine: chloroquine does not act in the liver on the exoerrhrocytic (hypnozoite) stage of *P. vivax* and *P. ovale.* The liver becomes a sanctuary and a source of intermittent reinfection despite adequate treatment with chloroquine. Primaquine, an 8-aminoquinoline tissue schizonticide active in the liver, effects radical cure thus preventing hypnozoite-induced relapses. It is usually recommended only after the patient has left the malaria zone. Malaria acquired by vertical transmission or blood transfusion does not require primaquine therapy because hepatic dormancy does not occur. Toxicity: Frequent - GI disturbance, hemolysis in G6PD deficiency. Raremethemoglobinemia in G6PD or nicotinamide-adenine dinucleotide (NAD) methemoglobin reductase deficiencies.

Quinine, quinidine- cinchona alkaloids. Quinine is the drug of choice in severe malaria or for multiresistant *P. falciparum.* The ora route is preferred for mild disease and i.v. for more severe disease. i.m. administration has good bioavailability but is painful and causes sterile abscesses. In the USA parenteral quinine is not readily available, causing delays in commencing treatment. Intravenous quinidine gluconate (an antiarrhythmic), the dextrorotatory diastereoisomer of quinine is widely available and of proven efficacy in severe malaria. It is now recommended by the CDC which has discontinued its supply of parenteral quinine. Intramuscular and oral quinidine have not been studied.

Quinine toxicity (ideal plasma concentration 8-20 mg/1): Frequent - cinchonism occurs at therapeutic blood levels and is not a reason for ceasing treatment. Hypoglycemia, a complication of malaria per se, may also be due to quinine stimulation of insulin production. It is *more* pronounced in the malnourished. Occasional - diarrhea, constipation, urticaria, bronchospasm, angioedema, hemolysis, agranulocytosis. Repeated courses may predispose to blackwater fever. ECG changes include prolonged QTc and T wave flattening. Rapid infusion causes hypotension, conduction disturbances, convulsions, deafness and death.

TABLE 9.3. DRUG TREATMENT OF PROVED OR SUSPECTED MALARIA

P. vivax, P. malaria, P. ovale, *sensitive* P. falciparum

Oral: chloroquine base 10 mg/kg (max. 600 mg), then 5 mg/kg/dose at 6, 24 and 48 h Parenteral: i.m. or s.c. chloroquine base 3.5 mg/kg/6 h; change to oral as soon as possible; total dose 25 mg/kg; i.v. chloroquine base by infusion 0.83 mg/kg/h for 30 h in 0.9% NaCl or 5% dextrose

Sick child or potential for resistant P. falciparum

Oral or nasogastric tube

Quinine sulphate 10 mg salt/kg/8 h for 3-7 days' *plus one of*

Fansidar single dose: < 1 year 1/4 tablet; 1-3 years 1/2 tablet; 4-8 years 1 tablet; 9-14 years 2 tablets; > 14 years 3 tablets, *or*

Tetracycline 5 mg/kg (max. 250 mg)/6 h x 7 days, *or*

Clindamycin 10-13 mg/kg/8 hx3 days, *or*

Mefloquine 25 mg/kg (max. 1,250 mg) once or given in 2 doses of 15 and 10 mg/kg 6 h apart; Toxic effects are common

Halofantrine 8 mg/kg/6 h x 3 doses; repeat after 1 week

Parenteral

Quinine dihydrochloride: intravenous - 7 mg salt/kg loading dose' i.v. over 30 min then 10 mg/kg in 10 mg/kg 5% dextrose over 4 h, and repeated every 8 h until child can tolerate oral therapy as above for total 7 days;

intramuscular - 20 mg/kg loading dose (divided sites) then 10 mg/kg/8 h changing to oral therapy as above for total 7 days, *or*

Quinidine gluconate: intravenous - 10 mg salt/kg (max. 600 mg) in 0.9% NaCl over I h, then continuous infusion of 0.02 mg/kg/min for max 3 days; *or* 15 mg/kg loading dose i.v. over 4 h then 7.5 mg/kg/8 h infused over 4 h; change to oral therapy as soon as possible as above for total 7 days

Radical cure of P. vivax *and P.* ovale *after leaving malaria zone*

Primaquine phosphate - 0.3 mg/kg/day (max. 15 mg) × 14 days; in SE Asia and Oceania, treat for 21 days; mild variants of G6PD deficiency may try 0.6 mg/kg/weekx 8 weeks monitoring hematocrit; screen for G6PD deficiency before use

---

Quinidine toxicity (ideal plasma concentration 5-10 mg/1): Compared to quinine it causes less cinchonism, more diarrhea, a fourfold greater effect in prolonging QTc and is more likely to induce hypotension.

Mefloquine: active against all malaria species including almost all multiresistant *P. falciparum.* Naturally resistant parasite populations exist. There is no parenteral formulation hence it is not advised in severe malaria. If given after quinine or chloroquine, allow 8 h to elapse to avoid cardiotoxicity. Toxicity: In treatment doses, side effects are common (see above).

Halofantrine: A phenanthrene-methanol compound active against multiresistant *P. falciparum.* There is no parenteral preparation. Oral bioavailablity is unreliable. Unsuitable for severe disease.

Pyronaridine: a naphthyridine derivative widely used in China is highly active against multiresistant *P. falciparum.* Oral bioavailability is poor. Intramuscular absorption is excellent.

Sesquiterpene lactones (artemisinine, artesunate, artemether): Artemisinine, a trioxane compound derived from the plant*Artemisia annua* (quingahaosu), is a rapidly acting, effective blood schizonticide used in traditional Chinese medicine. Artensunate, the sodium succinyl salt is water soluble but unstable in solution. Artemether is an oil-soluble methyl ether suitable for i.m. injection. These drugs lower fever and clear parasites more rapidly than quinine. Because of high recurrence rates and synergism in vivo with other antimalarials, they should not be used alone. There are no reports of significant toxicity. More clinical pharmacology studies are eagerly awaited before these drugs pass into common use outside China

Supportive Measures for Severe Malaria

If possible manage in an intensive care setting. Pulmonary edema and renal failure are uncommon in children. Nevertheless, fluid management should be extremely cautious, where possible adjusted by central venous pressure readings especially in the presence of hypotension or oliguria. Hypotension refractory to intravascular volume expansion requires inotropic agents and a diligent search for sepsis. Coagulopathy may occur secondary to thrombocytopenia, disseminated intravascular coagulation, sepsis or liver failure.

Convulsions: exclude hypoglycemia. Control seizures aggressively. Prophylactic phenytoin or phenobarbitone is advocated by many.

Hypoglycemia: should be preempted by frequent checks of blood sugar and infusion of 5% or higher concentrations of dextrose.

Severe anemia: chronic anemia is common in malarious areas and is well tolerated. In severe acute malaria the hematocrit can be expected to fall in the first 24 h by as much as the percentage of parasitized cells. In general, transfusion is indicated for hematocrit < 15%. Fresh blood is preferable. Choosing an elderly relative as a donor decreases the risk of HIV in areas where screening is unavailable.

Exchange transfusion: probably reduces morbidity and mortality of severe disease but in some countries the risks of the procedure (hypotension, HIV, hepatitis B and C), are more dangerous than the disease. It is unlikely to have a significant effect on parasitised erythrocytes in cerebral capillaries but it probably removes mediators such as TNF and red cell and parasite debris. It reduces parasite load, partially restores clotting factors, and safely corrects anemia in volume overloaded patients. Indications for partial or complete exchange are uncomplicated parasitemia of > 50%, or > 10% in those with cerebral, pulmonary, renal or hemostatic complications.

Treatments No Longer Recommended for Severe Malaria

These include corticosteroids, nonsteroidal antiinflammatory agents, anti-cerebral edema drugs (mannitol, urea), low molecular weight dextran, adrenaline, heparin, prostacyclin, oxpentifylline (Trental), hyperbaric oxygen, cyclosporin A, hyperimmune serum.

Other Complications of Malaria

Splenic rupture - usually treated with splenectomy. In adults a successful nonoperative approach is being used with repeated blood transfusions and strict bed rest for a week.

Tropical splenomegaly (hyperreactive malarial splenomegaly) - an abnormal immunological response to repeated malaria infection. Responds well to months-years of antimalarials given at prophylactic doses.

Quartan nephropathy - chronic *P. malariae* infection can cause immune complex deposition in the glomeruli causing nephrotic syndrome that responds poorly to antimalarials, steroids or cytotoxic drugs.

The Future

There is an ongoing search for effective vaccines, radically new antimalarial compounds, drugs that reverse chloroquine resistance (e.g. calcium antagonists), strategies to impede the progress of resistance and vector control.

LEISHMANIASIS

Visceral Leishmaniasis, Kala-azar (L. donovani)

Treatment: stibogluconate (Sb) sodium (a pentavalent antimony) 20 mg/ kg/day (max. 800 mg) i.v. or i.m. × 20 days (×30 days in Africa). The course may be repeated × 1-2. In Latin America meglumine antimonate is used instead of Sb at the above doses. Allopurinol or interferon-y (especially in AIDS) with Sb, have shown success, as has ketoconazole. Alternatives to the antimonials include pentamidine isethionate 2-4 mg/kg/day i.m. × 15 days or amphotericin B 0.5 mg/kg/day or 1.0 mg/kg every alternate day for 8 weeks. AIDS patients have fulminant disease with poor response to therapy.

Oriental Sore, Old World Cutaneous Leishmaniasis (L. tropica, L. major)

The decision whether or not to treat depends on the location and extent of the lesion. Standard treatment is Sb 10-20 mg/kg i.v. or i.m. for 10-30 days. Controversial alternatives are pentamidine 2-4 mg/kg 1-2 × week, amphotericin as above, intralesional interferon-y, topical paromomycin with methyl-benzethonium chloride, topical heat of 39-42 °C for 2 h/day over 12 days, curretage and others. In adults with *L. major,* ketoconazole 400 mg/day for 4-8 weeks has been effective.

American or New World Leishmaniasis (L. braziliensis, L. mexicana)

Use antimonials as per kala-azar. For mucous membrane involvement treat for 4 or more weeks. Standard alternatives include amphotericin B or pentamidine as above. Recently allopurinol has been shown to be as effective as Sb in *L. bruceipanamensis*.

VACCINATION

Traditional 'leishmanisation' involves exposing the buttocks of children (an inconspicuous site) to sandfly bites or scratching the skin with material from active lesions. After resolution of the sores, the child has immunity against reinfection. This technique predated Jenner by several centuries. Various modifications using killed or live cultured promastigotes do not have wide acceptance. Immunotherapy using BCG vaccine with killed leishmania promastigotes has been shown to be as effective and less toxic than antimony. There are probably cross-reacting epitopes between mycobacterial and leishmanial antigens.

Cytokines (mainly interferon-y released by antigen-specific CD4+ T cells enhance the antileishmanial activity of macrophages. Antigen-specific CD8+ T cells also aid the resolution of cutaneous and visceral leishmaniasis. Future vaccine development may concentrate on inducing the right antigen-specific CD4+ and CD8+ subsets.

TRYPANOSOMIASIS

Chagas Disease, American Trypanosomiasis (T. cruzi)

Current therapy is unsatisfactory. Only 2 drugs are of some value in the acute phase. Nifurtimox - 4-5 mg/kg/6 h (1-10 years), 3-4 mg/kg/6 h (11-16 years)x90-120 days. Cures a minority of patients but reduces mortality and morbidity in most. The addition of interferon-γ for 20 days may further shorten the illness. *Or* Benznidazole - 5-10 mg/kg/day × 60 days.

Supportive care: No drug is of value in the indeterminate or chronic (chagasic) phases. Consider anti-heart failure treatment for chagasic heart disease, cardiac pacing for bradyarrythmias, pneumatic dilation of the gastroesophageal junction for megaesophagus. Myotomies or esophageal resection with interposition of a loop of bowel are complicated by severe reflux.

Prevention: Gentian voilet added to banked blood slowly inactivates parasites. Mass screening of donated blood is not usually available. Large scale vector control programs and housing improvement are distant goals.

African Trypanosomiasis, Sleeping Sickness (*T. brucei gambiense*, *T. brucei rhodesiense*)

Hemolymphatic stage: suramin 20 mg/kg on days 1, *3, 7* 14 and 21 by slow 1.v. infusion of a 10% solution. Alternative - pentamidine 4 mg/kg/day i.m. × 10 days. For CNS involvement-melarsoprol, a highly toxic arsenical that crosses the blood-brain barrier. Use 0.36

mg/kg/dose i.v. increasing gradually to 3.6 mg/kg at intervals of 1-5 days for total of 10 doses over 1 month, for a total dose of 18-25 mg/kg. For both hemolymphatic and CNS disease eflornithine has shown great promise. Effectiveness in *T. brucei rhodesiense* has been variable. Dose: 100 mg/kg/6 h i.v. × 14 days, then 75 mg/ kg/6 h oral × 3-4 weeks.

TOXOPLASMOSIS

Prevention

Pregnant women should avoid ingestion of cysts or sporulated oocysts, i.e. cook meat well, wash fruit and vegetables, wear gloves when working in the garden, avoid contact with cat litter. Systematic serological screening of pregnant women can identify pregnancies at risk, however further investigation using fetal blood, amniotic fluid and brain ultrasound is needed to determine which fetuses are affected so that decisions regarding chemotherapy or abortion can be made. For infection during pregnancy spiramycin and/or pyrimethamine/ sulfadiazine are used.

Treatment

Congenital Toxoplasmosis. Infected children with or without signs should be treated with protocols (a) and (b) below. For those with overt disease, treatment should be for 1 year, starting with 6 months of (a) and during the following 6 months alternating monthly with (b). For subclinical disease treat with (a) for 6 weeks then commence a regimen of (b) for 6 weeks alternating with (a) for 4 weeks for a total of 1 year. In a healthy child awaiting diagnosis, with maternal infection proved to have been acquired in pregnancy treat with (a) for 1 month. The decision to continue treatment depends on serology and clinical findings in the infant, and requires expert consultation. In a healthy infant whose mother has a high Sabin-Feldman dye test titer but in whom the date of infection cannot be determined treat for 1 month with (b) and consult with an expert in the field. For chorioretinitis involving the macula, or for CNS involvement with CSF protein > 100 mg% before I month of age, add prednisone 1.5-2 mg/kg/day. Continue until the inflammatory process (e.g. high CSF protein) has abated. Pyrimethamine and the sulfonamides sequentially inhibit folic acid synthesis, resulting in dose-related bone marrow suppression.

Folinic acid minimises this risk without inhibiting the drug action on the trophozoites.

(a) Pyrimethamine 15 mg/m^2/day or 1 mg/kg/day (max 25 mg). For severe organ involvement, give loading dose of 2 mg/kg/dayx 3 days.

Plus sulfadiazine *or* trisulfapyrimidines (sulfadimidine, sulfamerazine, sulfadiazine) 42.5 mg/kg/ 12 h.

Plus folinic acid 5 mg orally every alternate day (i.m. in small infants). (b) Spiramycin 100 mg/kg/day orally in 2-3 doses.

Immunodeficient Patients. Toxoplasma encephalitis is seen especially in AIDS, and organ transplants. Other sites include the eye, lung, testes, blood. Use pyrimethamine 2 mg/kg/day × 3 days then 1 mg/kg/day (max. 25 mg) for 4 weeks.

Plus sulfadiazine or trisulfapyrimidines 100-200 mg/kg/day × 3-4 weeks.

Plus folinic acid 10 mg/day i.v., i.m., p.o. In AIDS, treatment should be lifelong because of high incidence of relapse on cessation of therapy.

For sulfa sensitivity use desensitization methods, change sulfonamide, or change to clindamycin (also useful in chorioretinitis). Cotrimoxazole and spiramycin are inferior drugs and should not be used. Fansidar has been effective in a biweekly regimen but interest has waned because of occasional relapses and Stevens-Johnson syndrome.

Investigational drugs include macrolides (roxithromycin, azithromycin, clarithomycin), folic acid antagonists (piritrexim, trimetrexate), and immunomodulators (interferon-γ, recombinant interleukin-2).

PNEUMOCYSTIS CARINII PNEUMONIA (PCP)

PCP is usually an opportunistic infection. Those with HIV infection are at special risk and therefore this section is devoted largely to them. Before commencing treatment seek a definitive diagnosis with bronchoalveolar lavage, open lung or transbronchial biopsy. The major drugs used are trimethoprim-sulfamethoxazole (TMP-SMZ), and pentamidine. Both are tolerated less well than usual in AIDS patients.

Treatment

(1) 20 mg TMP - 100 mg SMZ/kg/day in 3-4 divided doses i.v. Change to oral (same dose) after clear improvement for total of 2-3 weeks. If no improvement within 5-7 days change to pentamidine.

(2) Pentamidine isethionate - if TMP-SMZ fails or is not tolerated. 4 mg/kg once daily i.v. or i.m. × 2-3 weeks.

(3) Corticosteroids are recommended for those > 13 years with moderate-severe PCP preferably within 72 h of onset of symptoms.

TABLE 9.4. GUIDELINES FOR PCP PROPHYLAXIS IN HIV SEROPOSITIVE OR INFECTED CHILDREN

Previous PCP in any age
Age 1-12 months
CD4+ count < 1,500/mm³ or unknown, or
infected mother but child's serostatus and CD4+ unknown, or
Features of HIV (persistent thrush, failure to thrive, encephalopathy) with status of mother and baby unknown
12-24 months
CD4+ < 750/mm³ or < 20% of total lymphocyte count (TLC)
25 months to 6 years
CD4+ < 500/mm³ or < 20% of TLC
> 6 years
CD4+ count < 200/mm³ or < 20% of TLC

Other drugs with potential that are not yet recommended for children include TMP-dapsone, trimetrexate with folinic acid, pyrimethamine-sulfadoxine, aerosolized pentamidine, clindamycin with primaquine, and hydroxytlaphthquinone (566C80).

Prophylaxis in HIV Infection

Without prophylaxis, about 50% of children with AIDS develop PCP. The risk factors are previous PCP and a low CD4+ (T helper lymphocyte) count. Recent guidelines for prophylaxis take into account the higher ageappropriate norms for CD4+ in infants (table 4).

Recommended prophylaxis: TMP-SMX 150 mg TMP/m²/day+SMX 750 mg/m²/day in 2 divided dosesx3/week given on consecutive or alternate days. Some recommend giving the same daily dose 7 days/week. If TMP-SMX not tolerated try pentamidine (> 5 years) 300 mg/dose in 6 ml sterile water by Respiragard II inhaler (Marquest Medical Products, Englewood, Colo.), or dapsone (> I month) 6 mg/kg/week (max. 200 mg) orally. In those intolerant of inhaled pentamidine, i.v. pentamidine 4 mg/kg/dose every 2-4 weeks has been suggested.

Giardia lamblia Infection

Cysts or trophozoites can be detected by stool microscopy,

duodenal aspirate or biopsy, the duodenal string test, and recently by culture techniques. A commercial ELISA test now available correctly diagnoses many cases in which stool microscopy is negative. Avoid empirical trials of antigiardia therapy in suspected but unconfirmed giardiasis. Treatment of asymptomatic carriers is probably unnecessary as long term ill effects are negligible and conversion to a symptomatic state has not been described.

Treatment

Metronidazole and the single dose nitroimidazoles are effective, well tolerated and widely used. In the USA, quinacrine, an acridine dye, is still recommended. Furazolidone, a nitrofuran less effective but better tolerated than quinacrine, is available as a liquid.

Tinidazole - 50 mg/kg (max. 2 g) once, or

Ornidazole - 40 mg/kg (max. 1.5 g) once, or

Metronidazole - 5-8 mg/kg (max. 250 mg)/8h × 5-10 days, or Quinacrine - 2 mg/kg/8 h (max. 100 mg) × 5 days, or Furazolidone - 1.5 mg/kg/6 h × 7-10 days.

Failed Treatment

Try metronidazole with quinacrine when repeated courses of either drug have failed. Consider immunodeficiency (especially IgA), reinfection, lactose intolerance, other pathogens, and other diseases such as celiac disease. For suspected poor compliance use single dose tinidazole or ornidazole.

Prevention

Cysts can withstand standard chlorination of water. Travelers should boil drinking water for at least 1 minute. Alternatives include halazone (5 tablets/l/30 min) or a solution of crystalline iodine (12.5 ml/l/30 min) or chlorine (4-6 mg/1). Avoid uncooked foods, especially salads that may have been prepared with contaminated water.

BABESIA

This malaria-like zoonosis is rare and usually subclinical except in asplenic or immunocompromised individuals. Diagnosis is by microscopic examination of Giemsa or Wright stained thick and thin blood films.

Treatment: Clindamycin 7-13 mg/kg (max 600mg)/8 h i.v/i.m. x 7 days PLUS quinine 8-9 mg/kg/8 h (max 600 mg) × 7 days p.o., i.v., i.m. Alternatives: pentamidine with TMP-SMZ, the antitrypanosomal drug diminazene, or exchange transfusion.

Cryptosporidium

A coccidian parasite known to veterinarians for decades as a cause of diarrhea in young livestock and respiratory disease in birds. Since the early 1980s it has been recognized as causing self-limited diarrhea in the immunocompetent and severe chronic diarrhea in AIDS and other immune deficiencies. Transmission is usually via the fecal-oral route. It is primarily a zoonosis although person-to-person transmission is well described. Breastfeeding may be protective.

Treatment: There is no satisfactory anticryptosporidial agent despite a desparate need for such a drug in AIDS sufferers. Antidiarrheal agents are generally ineffective. The mainstay of treatment is good fluid and electrolyte therapy occasionally supported by parenteral nutrition. The macrolide spiramycin despite some anecdotal success does not seem to be effective. There are promising reports on the use of bovine hyperimmune colostrum, a dialysable calf lymph node extract, and eflornithine. One leukemic child improved dramatically after oral human immunoglobulin.

Isospora belli

Isosporiasis responds promptly to TMP-SMZ as per pneumocystis for 10 days. In AIDS, hypersensitivity to TMP-SMZ is common in which case pyrimethamine 50-75 mg/day may be used. Because of its high recurrence rate, AIDS patients should be kept on long-term, low-dose prophylactic TMP-SMZ (or Fansidar- < 4 years 1 /4 tablet/ week, 4-8 years 1 /2 tablet/ week; 9-14 years 3/4 tablet/week; 14 years I tablet/week).

Blastocytosis hominis and Dientamoeba fragilis

Their role in diarrhea and other gut symptoms as human pathogens or innocent bystanders is still debated. Treat if symptoms persist after other pathogens have been excluded. For *Blastocytosis* use metronidazole 15 mg/kg (max. 750 mg)/8 hx 10 days. For *Dientamoeba* treat as per giardiasis or use tetracycline 10 mg/kg/6 h for 10 days.

Balantidium coli

Usually asymptomatic but can cause a wide range of clinical problems including severe colitis and gut perforation. Treat with metronidazole 15 mg/ kg/8 hx 5 days or tetracycline 10 mg/kg (max. 500 mg)/6 hx 10 days, or iodoquinol 13 mg/kg/8 hx20 days.

Helminths

For those who have read of or seen how old antihelminth agents caused more death and misery than the disease they treated, the advent of safe broad spectrum antihelminthics over the last 1-2 decades seems almost miraculous. Many can b:, administered in a single dose or in a single day.

The benzimidazoles include thiabendazole, mebendazole, albendazole. They are useful against the nematode geohelminths (ascaris, trichuris, hookworm) and enterobius. Albendazole is also effective against strongyloides if used for 3 days, and echinococcus. There is hardly any indication today for thiabendazole that cannot be treated more comfortably with albendazole.

Praziquantel has revolutionized the treatment of schistosomiasis. It is also highly effective against the other trematode or fluke infections (lung, liver, intestinal), excepting *Fasciola hepatica* which is rare. It is useful against cestode (tapeworm) infections in conventional doses and cysticercosis and *Echinococcus* in high doses.

Ivermectin is a broad-spectrum veterinary antihelminth that has recently become the drug of choice for onchocerciasis (river blindness). It is effective against other nematodes including strongyloides (single dose) and the geohelminths except in hookworm.

Of the major worm infections in man, the guinea worms, *Fasciola trichinella,* hydatid cysts and cysticercosis remain significant therapeutic problems. The last three are probably responsive to albendazole and praziquantel either singly or combined.

The advent of safe broad-spectrum single-dose antihelminthics has led to the exciting prospect of mass treatment at regular (probably yearly) intervals in developing countries aimed at maintaining worm burdens below disease inducing thresholds rather than cure.

ECTOPARASITES

Scabies (Sarcoptes scabei)

Permethrin 1 % cream is the drug of choice. Apply to entire body and to all family members. Repeat after 1 week. Symptomatic improvement may take several weeks. Alternatives include Lindane 1 % - apply for 6 h, repeat after I week. Rarely causes neurotoxicity in children < 3 years. Benzoyl benzoate 10% is a malodorous cutaneous irritant, and is not the best drug. Crotamiton 10% applied twice daily for 5 days is less effective than Lindane.

Lice (Pediculosis capitus, humanus *and* Phthirus pubis)

For head and pubic lice use Lindane 1 % shampoo or permethrin 1 % hair cream. Repeat after 1 week. Soak combs in a pediculocide. Remove nits with a fine tooth-comb or forceps. For infestation of the eyelids, use pertolatumx 5 daily to asphyxiate the nits.

10

Enterobacteriaceae

Organisms in the family *Enterobacteriaceae* are facultative gram-negative rods that ferment glucose. They are oxidase negative, reduce nitrates to nitrites (except some strains of the genera *Enterobacter, Erwinia, Klebsiella, Xenorhabdus,* and *Yersinia),* and do not require nor is their growth enhanced by NaCl. They may be motile by peritrichous flagella (except *Tatumella)* or nonmotile. These characteristics serve to differentiate members of the family *Enterobacteriaceae* from other glucose-fermenting, facultative gram-negative rods which may fall in the family *Vibrionaceae* or *Pasteurellaceae*. Negative oxidase and/ or lack of NaCl requirement separate the *Enterobacteriaceae* from the *Vibrionaceae,* and negative oxidase and/or lack of requirement for organic nitrogen sources distinguish the *Enterobacteriaceae* from the *Pasteurellaceae*.

Enterobacteriaceae are widely distributed on plants, in soil, and in the intestine of humans and animals. They are associated with many types of human infections, including abscesses, pneumonia, meningitis, septicemia, and intestinal, urinary, and wound infections. *Enterobacteriaceae* are a major component of the normal human intestinal flora, but they are relatively uncommon in the normal flora of other body sites. Hospitalized patients often become colonized with *Enterobacteriaceae,* and these organisms are a major cause of nosocomial infections. *Enterobacteriaceae* may account for 80% of clinically significant isolates of gramnegative bacilli in clinical microbiology laboratories, and they may account for 50% of all clinically significant isolates. These bacteria are particularly common in certain types of specimens. For example, they account for nearly 50% of septicemia cases, 60 to 70% of bacterial enteritis cases, and more than 70% of urinary tract infections.

Since the last edition of this Manual, a number of changes have occurred in the nomenclature and classification of members of the family *Enterobacteriaceae*. The family now contains over 20 genera and more than 100 species. Recent changes in, and additions to, the family have been based on a multiphasic approach to taxonomy centered in genetic relatedness studies by DNA-DNA hybridization and tempered by phenotypic characteristics, including biochemical features and antimicrobial susceptibility. Many of the newly recognized genera and species are of no or unknown clinical significance and will very rarely be encountered in clinical microbiology laboratories. Approximately 49 species or groups are currently recognized as definite or probable human pathogens. The reader is referred to other sources on the *Enterobacteriaceae* for discussion of other species or groups that are not known to be human pathogens, and it should be recognized that any of these organisms may infect humans in selected circumstances.

INFECTIONS

Intestinal Infections

Members of the *Enterobacteriaceae* account for a major portion of intestinal infections in humans and animals in the United States and throughout the world. Although many of the *Enterobacteriaceae* have been implicated as a cause of diarrhea, only members of the genera *Escherichia, Salmonella, Shigella,* and *Yersinia* are clearly established as enteric pathogens.

Escherdchia coil. *E. coli* is associated with at least four types of human enteric disease, including enteropathogenic, enterotoxigenic, enteroinvasive, and hemorrhagic. Enteropathogenic strains cause diarrhea, mostly in infants, by unknown mechanisms. These organisms are recognized by serological testing, and they belong to a limited group of specific serotypes. Although many of these strains produce no known factors of pathogenesis, they are capable of causing diarrhea in normal volunteers. The occurrence of diarrhea due to enteropathogenic *E. coli* seems to have decreased markedly over the past 20 years.

Enterotoxigenic strains of *E. coli* cause secretory diarrhea by the elaboration of heat-labile or heatstable enterotoxins or both. Surface attachment factors are also important and appear to be required for the organisms to attach to the intestinal mucosa, where they then elaborate toxin(s). Enterotoxigenic *E. coli* strains typically cause profuse, watery diarrhea, and they are often implicated in cases of traveler's diarrhea. The importance of enterotoxigenic *E. coli* in diarrheal disease is largely unknown due to the lack of practical methods for enterotoxin detection.

Enteroinvasive strains of *E. coli* cause a dysenterylike illness similar to *Shigella* infection. These *E. coli* strains invade the mucosal cells, causing destruction of the cells followed by sloughing of areas of the mucosal lining. Typically, patients with diarrhea due to enteroinvasive *E. coli* have stools containing blood, mucus, and polymorphonuclear leukocytes. Enteroinvasive strains of *E. coli* are thought to occur less commonly than enterotoxigenic strains, but the true importance of enteroinvasive *E. coli* in the United States is unknown because a suitable assay for use in clinical laboratories to detect these strains has not been available. The recent association of a highmolecular-weight plasmid with invasiveness may provide a useful marker for the detection of these organisms.

Hemorrhagic colitis is a recently recognized enteric infection due to *E. coli* strains of a specific serotype, 0157:H7. These strains cause a severe diarrhea characterized by grossly bloody stools . Most of the reported cases have been acquired from the consumption of partially cooked hamburger. *E. coli* strains associated with hemorrhagic colitis produce a cytotoxin for Vero and HeLa cells that appears to be identical to the shiga toxin of *Shigella dysenteriae*. It is believed that this toxin damages vascular endothelial cells, resulting in hemorrhage into the bowel lumen.

Salmonella

Salmonellae cause a wide range of human enteric disease, from self-limited gastroenteritis with mild symptoms of short duration, to severe gastroenteritis with or without bacteremia, to typhoid fever, which is a severe, debilitating, and potentially life-threatening illness. *S. choleraesuis, S. paratyphi A,* and S. *typhi* are particularly important because of their frequent association with severe disease and bacteremia. Other *Salmonella* serotypes may also cause bacteremia, but they are more likely to cause gastroenteritis without bacteremia. Salmonellae typically invade through the bowel mucosa and multiply in the submucosa. Depending on the virulence of particular strains and the host response, they may invade and multiply in the bloodstream or the lymphatic tissue or both. Salmonellae are the most common cause of bacterial diarrhea in the United States.

Shigella

Shigella species cause classical bacillary dysentery characterized by severe, cramping abdominal pain and diarrhea with blood and mucus. These organisms invade mucosal cells, causing death and sloughing of the cells into the bowel lumen. However, they seldom invade beyond the mucosa. *Shigella* infections remain one of the most commonly recognized causes of bacterial diarrhea in the country.

Yersinia

Over the past 10 years, *YY enterocolitica* has been clearly demonstrated to be a significant cause of gastroenteritis . It is more often recognized in northern than southern parts of the northern hemisphere and is a major cause of enteric illness in most of states. Although the organism may produce an enterotoxin that closely resembles that of E. *coli* ST, *Y. enterocolitica is* considered primarily an invasive pathogen, and involvement of intestinal lymph nodes is common. Such infections are difficult to distinguish from acute appendicitis. It appears that, like *E. coli,* only strains of *Y. enterocolitica* that possess essential virulence factors are capable of causing human intestinal disease.

Extraintestinal Infections

Except for *Shigella* species, which rarely cause infections outside the gastrointestinal tract, most of the *Enterobacteriaceae* are capable of causing a variety of extraintestinal infections. However, a small number of species, including *Enterobacter aerogenes, E. cloacae, Escherichia coli, Klebsiella pneumoniae, Proteus mirabilis,* and *Serratia marcescens,* account for the vast majority of infections. Urinary tract infections, primarily cystitis, are the most common, followed by respiratory, wound, bloodstream, and central nervous system infections. Urinary tract, respiratory, and wound infections due *to Enterobacteriaceae* are each encountered more commonly than intestinal infections in most clinical microbiology laboratories. Many of these infections, especially sepsis and meningitis, are life threatening, and many infections of any type due to *Enterobacteriaceae* are hospital acquired. Because of the severe nature of many of these infections, prompt isolation, identification, and susceptibility testing of *Enterobacteriaceae* are essential.

COLLECTION OF SPECIMEN

Enterobacteriaceae may be recovered from infectious processes in virtually any body site, and good practices for specimen collection in general should be followed for the collection of blood, body fluid,. respiratory, wound, and other specimens for the culture of *Enterobacteriaceae*. Urine specimens should be collected by methods that reduce the possibility of contamination of the specimen. Such methods include the clean-catch, mid-stream urine technique, catheterization under aseptic conditions, and in certain instances, suprapubic aspiration. Most *Enterobacteriaceae* multiply rapidly in media such as urine, and it is important to transport specimens to the laboratory as quickly as possible to avoid overgrowth of such organisms. This is especially true for urine specimens because of the need for quantitation, and specimens should be refrigerated unless they can be cultured within 2 h of collection. Alternatively, urine preservative systems may be used in situations where specimen transportation is often delayed. The importance of rapid processing of specimens after collection cannot be overemphasized, and inadequate specimen delivery systems may reduce the quality of otherwise well-collected specimens.

Stool specimens also require special attention in both collection and transportation. Specimens should be collected early in the course of illness when the causative agent is likely to be present in largest numbers. With the exception of typhoid fever, enteric pathogens are more difficult to detect as the course of illness progresses, and cultures collected in the recovery phase may often be negative. Freshly passed stool is much preferred over rectal swabs for the culture of *Enterobacteriaceae,* and stool provides the highest recovery of enteric pathogens. Rectal swab specimens may be used in selected circumstances, such as for infants or for mass screening in investigation of epidemics, but these specimens must be collected properly by sampling the mucosa within the rectal vault to be of any value. Ideally, enteric specimens should be cultured within 2 h of collection, and if delays in processing are unavoidable, a transport medium may be used. Buffered glycerolsaline solution (pH 7.4) is often used for this purpose, but rapid processing rather than a transport medium should be used whenever possible.

CULTURE OF ENTEROBACTERIACEAE

Most *Enterobacteriaceae* grow readily on media commonly used in clinical microbiology laboratories, and specimens other than stool can be cultured on blood or chocolate agar if the specimens are from a normally sterile body site. Specimens such as urine and respiratory and wound samples that are likely to contain mixtures of organisms may be cultured on selective media to enhance the recovery of *Enterobacteriaceae*. MacConkey agar or eosin methylene blue agar may be used for this purpose, and these media offer the added advantage of differential features which allow a preliminary grouping of enteric and other gramnegative bacteria. In addition, broth enrichment cultures may be of value, especially when small numbers of *Enterobacteriaceae* may be present in the specimen.

Stool and rectal swab specimens require special culture procedures involving selective-differential media and enrichment broths. Selective media are those that allow the growth of *Enterobacteriaceae* while inhibiting the growth of normal stool flora organisms, and differential media are those that allow a preliminary characterization of organisms based on colony morphology and selected biochemical reactions. Media may be highly selective, allowing the growth of only a limited group of organisms (e.g., bismuth sulfite agar for *Salmonella),* or only moderately selective, allowing the growth of most gram-negative organisms and inhibiting predominantly gram-positive organisms (e.g., MacConkey agar). Selective-differential media are essential for the efficient recovery of enteric pathogens from stool specimens because these media reduce the overgrowth of pathogens by normal flora organisms and aid in the recognition of pathogens in primary cultures. Broth enrichments are also useful for the culture of stool specimens because they provide for the growth of enteric pathogens while retarding the growth of normal flora organisms. Such media are particularly useful when pathogens may be present in low numbers. In the acute phases of gastrointestinal illness, broth enrichments are probably not essential because enteric pathogens are usually present in high numbers, but specimens are often not submitted to the laboratory until the later stages of illness, and broth enrichment may be the only means of detecting the infecting organism in such cases. Broth enrichment may also be useful in epidemiological studies to enhance the detection of low numbers of organisms from asymptomatic or convalescing patients. Gram-negative broth and selenite broth are enrichment media that have been found useful in many laboratories through the years. Cold enrichment in phosphate-buffered saline has also been found to enhance the recovery of *Y. enterocolitica* from stool specimens, but this technique is apparently useful only in cultures of convalescent or asymptomatic subjects.

A battery of media for routine use in the culture of *Enterobacteriaceae* from stool specimens should include a nonselective medium such as blood agar, a moderately selective differential medium such as MacConkey agar, a more selective differential medium such as xylose-lysine-deoxycholate agar or Hektoen enteric agar, and a broth enrichment. The use of such a battery will provide for the effective detection of the *Enterobacteriaceae* most commonly associated with enteric infection, including *Salmonella* and *Shigella* species and *E. coli*. In areas where *Y. enterocolitica is* encountered, a selective-differential medium such as *Yersinia* selective agar may also be added. A complete stool culture procedure will also include media for the isolation of *Campylobacter* species and possibly media for the isolation of *Vibrio* species in areas where cholera and other *Vibrio* infections occur.

BIOCHEMICAL TESTING

Members of the family *Enterobacteriaceae* are identified primarily by testing for biochemical reactions. Several methods are available for the performance of biochemical tests, including a variety of commercial systems and the classical tubed media. In addition, various combinations of classical tests are available, and simple screening tests for identification have been proposed. Only the classical biochemical tests and rapid screening tests will be considered here because they are the basis for the taxonomy and identification of the *Enterobacteriaceae*.

TABLE 10.1. CLINICALLY SIGNIFICANT MEMBERS OF THE FAMILY *ENTEROBACTERIACEAE* AND CHANGES IN THEIR TAXONOMY AND NOMENCLATURE SINCE THE LAST EDITION OF THE MANUAL

Current Classification	*Previous Designation*
Cedecea davisae	None
C. neteri	None
Citrobacter freundii	Same
C. diversus	Same
C. amalonaticus	Same
Edwardsiella tarda	Same
Enterobacter cloacae	Same
E. aerogenes	Same

E. agglornerans	Same
E. gergoviae	Same
E. sakazakii	Same
Escherichia coli	Same
Ewingella americana	None
Hafnia alvei	Same
K. oxytoca	Same
K. ozaenae	Same
K. rhinoscleromatis	Same
K. planticola	None
Kluyvera ascorbata	None
K. cryocrescens	None
Morganella morganii	Same
Proteus mirabilis	Same
P. penneri	None
P. vulgaris	Same
Providencia stuartii	Same
P. rettgeri	Same
P. alcalifaciens	Same
Salmonella typhi	Same
S. choleraesuis	*Same*
S. paratyphi A	*S. enteritidis bioserotype* paratyphi A
"Arizona group"	*Arizona hishawii*
Salmonella serotype	*S. enteritidis* bioserotype
Serratia marcescens	Same
S. liquefaciens group	*S. liquefaciens*
S. odori fera	None
S. rubidaea	Same
Shigella dysenteriae	Same
S. flexneri	Same
S. boydii	Same
S. sonnei	Same
Tatumella ptyseos	None
Yersinia enterocolitica	Same
Y frederiksenii	Same

Y. intermedia	Same
Y. kristensenii	None
YY pestis	Same
YY pseudotuberculosis	Same

Classical Biochemical Tests

Of the many biochemical tests that are available, a battery of 28 tests is sufficient to differentiate all the currently recognized, clinically significant *Enterobacteriaceae. A* primary battery of 14 tests, including adonitol fermentation, arginine dihyrolase, citrate utilization, DNase, gas from glucose, hydrogen sulfide, indole production, lysine decarboxylase, motility, ornithine decarboxylase, phenylalanine deaminase, sucrose fermentation, urease, and Voges-Proskauer, is sufficient for the identification of most isolates to the genus level and of many isolates to the species level. *A* secondary battery of 14 additional tests may be used as a supplementary battery to identify more difficult isolates. The secondary test battery consists mainly of carbohydrate fermentation tests plus tests for the hydrolysis of ortho-nitrophenylgalactoside (ONPG) and for growth in KCN (because of the potential hazard in the preparation of the KCN medium, it is recommended that clinical laboratories purchase the medium from commercial suppliers). Several media are available for combined testing of the primary test battery reactions. Triple sugar iron agar provides results for gas and H2S production and also for the fermentation of glucose. Motility-indoleornithine medium provides results in a single medium for motility, indole production, and omithine decarboxylase. Phenylalanine-urease broth provides testing for phenylalanine deaminase and urease. Thus, the 14 primary test battery reactions can be obtained from 10 tubed biochemical media. However, each of the secondary test battery reactions must be performed individually. Clinically significant isolates that can not be identified by the primary and secondary test batteries should be sent to a reference laboratory for further testing.

Screening Tests

Test batteries that are rapid or abbreviated or both may be used to identify or aid in the identification of *Enterobacteriaceae. A* spot test for oxidase is an essential preliminary screening test that should be performed early in the identification of all isolates. Since all *Enterobacteriaceae* are oxidase negative, this simple test can save much unnecessary work by the elimination of oxidase-positive organisms.

However, caution should be exercised in the use of this test because false-negative reactions may occur with isolates tested from selective-differential media.

E. coli and swarming *Proteus* species make up a substantial percentage of the *Enterobacteriaceae* encountered in clinical laboratories, and these organisms can be accurately identified by experienced clinical microbiologists by colony morphology and a spot indole test. Flat, lactose-positive colonies surrounded by a zone of precipitated bile on MacConkey agar can be identified as *E. coli if* they give a positive spot indole test. For typical growth of swarming *Proteus,* spot indole-negative isolates can be identified as *P. mirabilis*. Many of the other commonly encountered organisms can be readily identified by using the 10 tubed media described above for the primary test battery.

IDENTIFICATION

Cedecea Species

Members of the genus *Cedecea* resemble *Serratia* species in many of the commonly tested biochemical characteristics, but *Cedecea* species are DNase negative. Members of this genus are also lipase positive (corn oil) and gelatin hydrolysis negative. Members of this genus are rarely recovered from clinical specimens, but when encountered, isolates of *C. davisae* and *C. neteri* may be of clinical significance. More than 90% of these strains give positive reactions for citrate utilization, motility, ONPG, and acid production from sucrose, malonate, mannitol, salicin, trehalose, and xylose. More than 90% of the strains are also negative for HS, indole, lysine decarboxylase, phenylalanine deaminase, and urease. The two potentially significant species are distinguished by sorbitol fermentation and ornithine decarboxylase. *C. davisae is* sorbitol negative and ornithine positive, whereas *C. neteri is* sorbitol positive and ornithine negative.

Citrobacter Species

Members of the genus *Citrobacter* typically give positive tests for citrate, gas from glucose fermentation, motility, ONPG, and fermentation of arabinose, mannitol, rhamnose, sorbitol, trehalose, and xylose. *Citrobacter* species are also negative for DNase, lysine decarboxylase, phenylalanine deaminase, Voges-Proskauer, and inositol fermentation. The three currently recognized *Citrobacter* species, *C. amalonaticus, C. diversus,* and *C. freundii,* are all wellrecognized human pathogens that cause a variety of infections. *C. freundii* strains often produce H_2S and may be confused with *Salmonella* species unless adequate biochemical testing is performed. *Citrobacter* isolates may also give false-positive agglutination tests in *Salmonella* grouping sera.

Adonitol, H_2S indole, KCN, and malonate are useful tests for the identification of *Citrobacter* isolates. *C. amalonaticus* isolates are adonitol, H_2S, and malonate negative but indole and KCN positive. *C. diversus* isolates are adonitol, indole, and malonate positive but H_2S and KCN negative. *C. freundii* isolates are adonitol, indole, and usually malonate negative.

Edwardsiella Species

E. tarda is the single clinically significant species of the genus *Edwardsiella,* and this organism is a wellestablished cause of a variety of extraintestinal human infections. The role of *E. tarda* as an intestinal pathogen is yet to be firmly established. Recently, new species and groups of the genus *Edwardsiella* have been described, but there is no evidence that they cause human infections. Clinically significant *Edwardsiella* strains are positive for gas from glucose fermentation, H_2S, indole, lysine decarboxylase, motility, and ornithine decarboxylase. They are negative for other biochemical tests commonly performed.

Enterobacter Species

The genus *Enterobacter is* a large and heterogeneous group of organisms. Most *Enterobacter* species are characterized by positive tests for citrate, gas from glucose, motility, ornithine decarboxylase, VogesProskauer, ONPG, and fermentation of sucrose, arabinose, mannitol, melibiose, raffmose, rhamnose, trehalose, and xylose. Most members of the genus are negati ve for DNase, H_2S, indole, and phenylala nine deaminase. Although there are eight species of *Enterobacter* currently recognized taxonomically, only five species are clinically significant in human infections. *E. aerogenes* and *E. cloacae* are the members of the genus most often encountered in clinical laboratories, and these are species clearly associated with pulmonary, bloodstream, central nervous system, soft tissue, genitourinary, and other extraintestinal infections of humans. *E. agglomerans is* also a well-established human pathogen that has particularly been associated with nosocomial infections. *E. gergoviae* and *E. sakazakii* are more recently recognized species. *E. gergoviae* has been associated with urinary tract, pulmonary, and bloodstream infections, and *E. sakazakii* has been associated with several types of infections, most notably with neonatal sepsis and meningitis.

TABLE 10.2: BIOCHEMICAL DIFFERENTIATION OF *CITROBACTER* SPECIES

Test	*Reaction of*		
	C. amalonaticus	***C diversus***	***C. freundii***
Adonitol fermentation	—	+	—
H_2S production (triple sugar iron)	—	—	+
Indole production	+	+	—
KCN, growth in	+	—	+
Malonate utilization	—	+	V

'Tests useful in the differentiation of *Enterobacter* species include adonitol and sorbitol fermentation, arginine dihydrolase, KCN, and lysine and ornithine decarboxylase. *E. aerogenes is* distinguished by positive reactions for all tests except arginine, and *E. agglomerans is* negative or variable for the reactions listed. This latter species is poorly defined taxonomically and consists of several genetically distinct groups as determined by DNA hybridization. This diversity makes exact biochemical identification difficult. For practical purposes, *Enterobacter* strains that are arginine, lysine, and ornithine negative can be reported as *E. agglomerans*. Most *E. agglomerans* isolates are also yellow pigmented, which may be useful in their identification. *E. cloacae* isolates are arginine, ornithine, KCN, and sorbitol positive, but lysine negative. Most isolates are also adonitol negative. *E. gergoviae is* lysine and ornithine positive, but negative for adonitol, arginine, KCN, and sorbitol. *E. sakazakii is* arginine, ornithine, and KCN positive, but adonitol, lysine, and sorbitol negative. This species is also yellow pigmented and gives a delayed-positive DNase test.

Escherichia Species

For many years, *E. coli* was the only species in the genus *Escherichia,* but four additional species are now recognized. Three of the four species, i.e., *E. fergusonii, E. hermannii,* and *E. vulneris,* have been isolated from potentially significant human sources, but at this time they are not yet proven to be human pathogens. It should also be recognized that the new *Escherichia* species are uncommon, and *E. coli* accounts for more than 99% *of Escherichia* isolates. *E. coli* is typically positive for gas production from glucose, indole, lysine, arabinose, mannitol, ONPG, trehalose, and xylose. Isolates are also generally negative for DNase, H_2S, phenylalanine deaminase, urease, Voges-Proskauer, inositol, and KCN. Other fermentation tests, not included in the recommended routine batteries, that may be useful in differentiating *E. coli* from the other species include cellobiose, arabitol, and mucate. *E. coli* is mannitol and mucate positive, but cellobiose and arabitol negative. These additional tests may be needed when dealing with isolates of *E. coli* from hemorrhagic colitis cases (see Serological Testing below) because these isolates are typically sorbitol negative.

TABLE 10.3: BIOCHEMICAL DIFFERENTIATION OF ENTEROBACTER SPECIES REACTION OF:

Test	*E aerogenes*	*E. agglomerans*	*E. cloacae*	*E. gergoviae*	*E sakazakii*
Adonitol fermentation	+	—	V	—	—
Argininedihydrolase	+	—	+	—	+
KCN, growth in	+	V	+	—	+
Lysine decarboxylase	+	—	—	+	—
Ornithin decarboxylase	+	—	+	+	+
Sorbitol fermentation	+	V	+	—	—

Ewingella Americana

The recently named genus *Ewingella* has a single species, *E. americana*. These organisms are typically citrate, VogesProskauer, and ONPG positive, and they ferment mannitol, salicin, and trehalose. *Ewingella* isolates are typically decarboxylase and phenylalanine deaminase negative, and they give negative or variable reactions for other commonly employed biochemical tests. Approximately onefourth of *E. americana* isolates studied to date have been from blood cultures.

Hafnia Alvei

The genus *Hafnia* contains a single species, *H. alvei*. Organisms classified as *H. alvei* were previously known as *"Enterobacter hafniae.* " *H. alvei is* characterized by positive reactions for lysine decarboxylase, motility, ornithine decarboxylase, growth in KCN, ONPG, and fermentation *of* arabinose, mannitol, rhamnose, trehalose, and xylose. *H. alvei* isolates are negative for adonitol, arginine, indole, and urease. Although infrequently encountered compared with other *Enterobacteriaceae, iso*lates *of H. alvei* may cause a variety *of* extraintestinal infections.

Klebsiella Species

Members *of* the genus *Klebsiella* give negative or variable reactions for most *of* the primary battery *of* biochemical tests. *Klebsiella* species are characteristically adonitol positive, but arginine, DNase, H_2S, motility, ornithine, and phenylalanine negative . They are also active in carbohydrate fermentations, giving positive tests for arabinose, mannitol, melibiose, raffmose, salicin, sorbitol, trehalose, and xylose. There are seven currently recognized *Klebsiella* species, but only five species are known to be clinically significant. Two of these five species, *K ozaenae* and *K rhinoscleromatis,* are variants of *K pneumoniae,* but they are considered separate species because of their association with specific human diseases. These five clinically significant species *(K. oxytoca, K. ozaenae, K. planticola, K. pneumoniae,* and *K. rhinoscleromatis)* can be distinguished by tests for indole, malonate, ONPG, and Voges-Proskauer. *K oxytoca is* distinguished by a positive indole reaction; *K ozaenae is* negative for malonate and Voges-Proskauer; *K pneumoniae is* indole negative, but malonate, ONPG, and Voges-Proskauer positive; and *K. rhinoscleromatis is* indole and ONPG negative. *K planticola,* the most recently recognized species, is similar to *K pneumoniae* or *K oxytoca* but can be distinguished from these species by its ability to ferment glucose at 5 but not at 45°C. In addition, an ornithine-positive biogroup of *K planticola* has been recognized, and isolates in this group are often clinically significant. *Klebsiella* species account for a significant proportion of urinary tract, bloodstream, and other extraintestinal infections. *K pneumoniae* and *K. oxytoca* are the most commonly encountered members of the genus, and although the other species are rarely isolated, they may cause significant infections.

Kluyvera Species

Kluyvera species are positive for ornithine decarboxylase, ONPG, and fermentation of arabinose, lactose, mannitol, melibiose, raffinose, rhamnose, salicin, trehalose, and xylose. There are two currently recognized species *of Kluyvera, i.e., K. ascorbata* and *K cryocrescens.* These species are very similar biochemically, and it is recommended that clinical laboratories report the genus identification without attempting to identify the species. However, the speciescan be separated by the ascorbate test, with *K. ascorbata* giving a positive test. In addition, *K ascorbata is* resistant to cephalothin and carbenicillin, whereas *K cryocrescens is* susceptible. *Kluyvera* isolates have occasionally been associated with significant human infections, and the two species seem to have equal pathogenic potential. However, *K ascorbata* appears to be numerically the more important of the two.

TABLE 10.4. BIOCHEMICAL DIFFERENTIATION *OF KLEBSIELLA* SPECIES

Test	Reaction of				
	K oxytoca	*K ozaenae*	*K plan-ticola*	*K pneu-moniae*	*K rhino sclero matis*
Glucose fermentation at 5°C	—	—	+	—	—
Indole production	+	—	V	—	—
Malonate utilization	+	—	+	+	+
ONPG	+	V	+	+	—
Voges-Proskauer	+	—	+	+	+

Morganella Morganii

The genus *Morganella* contains a single species, *M. morganii,* which was previously classified in the genus *Proteus. M. morganii is* a well-recognized human pathogen that has been associated primarily with urinary tract infections but also causes a variety of other human infections. The organism is characteristically positive for indole, motility, omithine decarboxylase, phenylalanine deaminase, urease, and growth in KCN. It is generally inactive in carbohydrate fermentations and rarely ferments sugars other than glucose. Tests for H_2S, indole, omithine, and xylose fermentation are useful for separating *M. morganii* from members of the genus *Proteus*.

Proteus Species

Members of the genus *Proteus* are positive for motility, phenylalanine deaminase, urease, growth in KCN, and xylose fermentation. *Proteus* species are also characteristically negative for adonitol, arginine, lysine, and most carbohydrate fermentations other than glucose and xylose. There are four currently recognized *Proteus* species, but only three, *P. mirabilis, P. penneri,* and *P. vulgaris,* are known to cause human disease. *Proteus* species are best known as urinary tract pathogens, but they are also well-documented agents of various systemic and localized extraintestinal infections. The three clinically significant *Proteus* species can be differentiated by tests for indole and ornithine decarboxylase. *P. mirabilis is* indole negative and ornithine positive; *P. penneri is* indole negative and ornithine negative; and *P. vulgaris is* indole positive and omithine negative. *P. penneri is* also resistant to chloramphenicol, whereas *P. mirabilis* and most *P. vulgaris* strains are susceptible.

Providencia Species

Clinically significant members of the genus *Providencia* are characterized by positive tests for citrate utilization, indole production, motility, phenylalanine deaminase, and growth in KCN.

TABLE 10.5. BIOCHEMICAL DIFFERENTIATION OF *PRO VIDENCIA* SPECIES

Test	*Reaction of:*		
	P. alcali-faciens	*P. rettgeri*	*P. stuartii*
Adonitol fermentation	+	+	—
Trehalose fermentation	—	—	+
Urea hydrolysis	—	+	*V*

Providencia species are also negative for arginine dihydrolase, DNase, H_2S, lysine and ornithine decarboxylases, VogesProskauer, and ONPG. They frequently fail to ferment carbohydrates other than glucose. Four species are currently recognized in the *Providencia* genus, but only three, i.e., *P. alcalifaciens, P. rettgeri,* and *P. stuartii,* are known to be human pathogens. These species are most commonly associated with urinary tract infections, but they may also cause a variety of other infections. The *Providencia* species can be distinguished biochemically by testing for adonitol fermentation, trehalose fermentation, and urease. *P. alcalifaciens is* adonitol positive, but trehalose and urease negative; *P. rettgeriis* adonitol and urease positive, but trehalose negative; and *P. stuartii is* adonitol negative, trehalose positive, and either positive or negative for urease.

Salmonella Species

Members of the genus *Salmonella* are causative agents of a variety of infections ranging from simple gastroenteritis to severe illnesses such as enteric fever with bacteremia. Typhoid fever is an example of the life-threatening illnesses that may be associated with members of the genus *Salmonella.* These organisms may also cause local infections of virtually any organ as a result of seeding during bacteremia.

Most *Salmonella* isolates are characterized by positive reactions for motility and fermentation of mannitol and sorbitol. They are also characteristically negative for DNase, indole, phenylalanine deaminase, urease, Voges-Proskauer, growth in KCN, ONPG, and fermentation of adonitol, sucrose, lactose, raffinose, and salicin, and they fail to utilize malonate. Most clinical *Salmonella* isolates are also H_2S positive. However, H_2S production is considered a variable characteristic for the genus overall because only *50%* of *S. choleraesuis* and 10% of S. *paratyphi A* are H_2S positive.

The classification of *Salmonella* species has been controversial for several years. Until recently, the Centers for Disease Control reported three biochemically distinct species of *Salmonella, i.e., S. typhi, S. choleraesuis,* and *S. enteritidis,* along with a separate genus for *"Arizona,* " and this approach was followed in the previous edition of this Manual. However, all salmonellae appear to be very closely related, and the Centers for Disease Control have now revised this approach to be more consistent with the traditional reporting of serotypes rather than species. Reporting of specific serotypes requires the availability of an extensive battery of antisera and the labor resources to use them. Most clinical laboratories are unable to identify *Salmonella* isolates to specific serotypes, and serogroup reporting is more common. Isolates are then referred to reference laboratories for specific serotyping. This approach is sufficient for most *Salmonella* isolates, but *S. typhi, S. choleraesuis,* and *S. paratyphi A* should be specifically identified in clinical laboratories because of their special clinical significance. Hence, in this edition of the Manual, we recommend that *Salmonella* isolates be identified biochemically as *S. choleraesuis, S. typhi,* or *S. paratyphi A,* with serogroup confirmation. Other *Salmonella* isolates should be confirmed by serogrouping and reported as a certain Salmonella serogroup, with specific serotypes to be reported subsequently, after reference laboratory testing. The final report for these isolates should include the genus and serotype name, e.g., *Salmonella* serotype *typhimurium* or, for simplicity, *S. typh imurium.*

S. choleraesuis, S. paratyphi A, and S. *typhi* can be separated by tests for arabinose fermentation, citrate utilization, gas from glucose, lysine decarboxylase, ornithine decarboxylase, rhamnose fermentation, and trehalose fermentation. *S. choleraesuis is* arabinose and trehalose negative, occasionally positive for citrate, and positive for the other tests. *S. paratyphi A* is characteristically citrate and lysine negative, but positive for the other reactions. *S. typhi* is positive only for lysine and trehalose, and other *Salmonella* serotypes are positive for all the tests. Isolates of *S. typhi* are also recognizable by their typical reactions in triple sugar iron agar including acid butt, alkaline slant, a band of H_2S formation at the slant-butt interface, and lack of gas production. Specific identification of *S. typhi, S. paratyphi A,* and *S. choleraesuis* should be accomplished as soon as possible after isolation in the clinical laboratory because of the propensity of these organisms to cause life-threatening infections of considerable epidemiological significance. It is also important to identify other *Salmonella* isolates at the genus and serogroup level, but specific serotype identification can await

TABLE 10.6. BIOCHEMICAL DIFFERENTIATION OF SELECTED MEMBERS OF *THE SALMONELLA* GROUP

	Reaction of			
Test	***S. cholerae-suis***	***S. para-typhi***	***S. A tvphi***	***n Other***
Arabinose fermentation	—	+	—	+
Citrate utilization	V	—	—	+
Glucose gas production	+	+	—	+
Lysine decarboxylase	+	—	+	+
Ornithine decarboxylase	+	+	—	+
Rhamnose fermentation	+	+	—	+
Trehalose fermentation	—	+	+	+

results from a reference laboratory. The identity of all *Salmonella* isolates, whether at the genus or species level, should be confirmed both biochemically and serologically (see below).

The "Arizona group" was once considered a separate genus, but these isolates are now known to be very closely related to strains of *Salmonella*. They are now considered members of this genus. These organisms will be reported by reference laboratories as *Salmonella* serotypes, e.g., *Salmonella* serotype 47:r:z.

Serratia Species

Members of the genus Serratia are characterized by negative reactions for arginine dihydrolase, H,S production, phenylalanine deaminase, and urease. Serratia isolates are unusual among the Enterobacteriaceae in their production of three extracellular hydrolytic enzymes, i.e., DNase, gelatinase, and lipase. In addition, members of the genus characteristically give positive reactions for citrate, motility, Voges-Proskauer, ONPG, and fermentation of mannitol and trehalose. Members of the genus originally were often red pigmented, but most strains isolated in clinical laboratories today are nonpigmented.

Three species and one group of the genus *Serratia* are currently recognized as potential human pathogens. *S. marcescens* has a colorful history of use in experiments during a time when it was thought to be a harmless commensal. However, it is now recognized as a frequent cause of extraintestinal human infections ranging from simple cystitis to life-threatening bloodstream and central nervous system infections. It is particularly noted as a cause of nosocomial infections, and many hospitals harbor multiply antibiotic-resistant strains of *S. marcescens*. *S. liquefaciens* was, until recently, considered a single distinct species. However, it is currently recognized as a collection of closely related species that cannot be readily separated in clinical laboratories. Therefore, these organisms should be reported as "*S. liquefaciens* group." Members of this group are only occasionally encountered in clinical laboratories in the United States, and they are rarely associated with human infections. *S. odorifera is* a recently recognized species that is rarely encountered in clinical laboratories but has been associated with central nervous system and bloodstream infections. *S. rubidaea is* a well-known species that is occasionally associated with significant human infections but is most often an insignificant isolate from sputum, wound, or stool cultures. At least three other *Serratia* species are currently recognized, but there is no evidence that they are of any clinical significance.

Differentiation of the clinically significant members of the *Serratia* genus can be accomplished by testing for ornithine decarboxylase and fermentation of adonitol, arabinose, and sorbitol . Members of the *S. liquefaciens* group are arabinose, ornithine, and sorbitol positive, but adonitol negative. *S. marcescens* is ornithine and sorbitol positive, but arabinose negative. *S. odorifera is* arabinose and sorbitol positive and variably positive for adonitol and ornithine. Differentiation of *S.*

odorifera from the *S. liquefaciens* group may be difficult biochemically, but *S. odorifera* has a characteristic pungent odor that has been variously described as musty, potatolike, or vegetablelike. *S. rubidaea is* identified by positive reactions for adonitol and arabinose, but negative reactions for ornithine and sorbitol.

Shigella Species

Members of the genus *Shigella* have long been recognized as causative agents for bacillary dysentery. Although these organisms are very important as a cause of gastrointestinal infections, they rarely cause other types of infections. *Shigella* species are genetically very closely related to *E. coli* and would be considered taxonomically as members of *the Escherichia* genus. However, because of their distinctive ability to cause dysentery, *Shigella* species should continue to be reported as a distinct genus in clinical laboratories. Members of the genus *Shigella are* among the biochemically least-reactive members of the *Enterobacteriaceae*. They are characteristically negative for adonitol, citrate, DNase, gas from glucose, H_2S, lysine, motility, phenylalanine, sucrose, urease, Voges-Proskauer, inositol, KCN, lactose, malonate, salicin, and xylose. Members of the genus are no more than variably positive for any of the commonly employed biochemical tests other than glucose fermentation and nitrate reduction.

TABLE 10.7. BIOCHEMICAL DIFFERENTIATION OF *SERRATIA* SPECIES

Test	*Reaction of*			
	S. liquefaciens group	*S. marcescens*	*S. odorifera*	*S. rubidaea*
Adonitol fermentation	—	V	V	+
Arabinose fermentation	+	—	+	+
Ornithine decarboxylase	+	+	V	—
Sorbitol fermentation	+	+	+	—

There are four currently recognized *Shigella* species, i.e., *S. dysenteriae, S. flexneri, S. boydii,* and *S. sonnei*. All are capable of causing dysentery, and *S. dysenteriae* has been associated with a particularly severe form of illness thought to be related to toxin production. The identification of *Shigella* species is important for both clinical and epidemiological implications. Three biochemical tests are helpful in the separation of the *Shigella* species. *S. dysenteriae is* mannitol and ornithine negative, and 40% of the isolates are ONPG positive. *S. flexneri* is mannitol positive, but ONPG and ornithine negative, and *S. boydii* has the same reactions except for occasional (10%) ONPG-positive strains.*S. sonnei* gives positive reactions for mannitol, ONPG *(90%)*, and ornithine and is the only *Shigella* species that is ornithine positive. Serological testing is also needed for the identification of *Shigella* isolates. Although *S. dysenteriae* and *S. sonnei* are biochemically distinct, *S. flexneri* and *S. boydii* are often biochemically identical. Serological grouping is essential for the separation of *S. flexneri* and *S. boydii* and is also required for the confirmation of *S. dysenteriae* and *S. sonnei*.

Tatumella Ptyseos

Genus *Tatumella* is a new genus with a single species, *T. ptyseos*, which was previously designated CDC group *EF-9*. Strains of *TT ptyseos* have been placed in the *Enterobacteriaceae*, based on DNA hybridization studies, despite some unusual characteristics, including susceptibility to penicillin and polar or lateral rather than peritrichous flagella. The organism has been recovered from a variety of clinical specimens, including blood cultures. Therefore, it is considered a rare but potentially significant human pathogen. Isolates of*T. ptyseos* are relatively inactive biochemically at 35 to 37°C but are more active at 25°C. They are typically negative for all of the commonly employed biochemical tests except phenylalanine deaminase and fermentation of sucrose and trehalose. Isolates may occasionally give positive reactions for fermentation of melibiose, raffinose, and salicin. A positive test for phenylalanine deaminase with negative indole and urease reactions is highly suggestive of *Tatumella*.

Yersinia Species

Members of the genus *Yersinia* are characterized by negative or variably positive reactions for most of the commonly used biochemical tests. All members of the genus are negative for arginine dihydrolase, citrate utilization, DNase, H_2S, motility, phenylalanine deaminase, Voges-Proskauer, and malonate utilization. They are also positive for fermentation of mannitol and trehalose. Some strains give additional positive reactions (especially for motility and VogesProskauer) if incubated at 25 instead of 35 to 37'C. Six clinically significant species of the genus *Yersinia* are currently recognized, and there are at least two other species and groups that will not be discussed here because they are of no known clinical significance.

It is well established that some strains of *Y. enterocolitica* can cause gastrointestinal infections, and they may cause a severe form of gastroenteritis that mimics appendicitis. These infections may occur as sporadic cases or as food- or waterborne outbreaks. The organism may also less commonly cause extraintestinal infections. The characteristics that separate *Y. enterocolitica* from other *Yersinia* species include positive tests for ornithine decarboxylase and sucrose fermentation and negative tests for melibiose, raffinose, and rhamnose fermentation. More than 70% of the isolates are also positive for urease.

Y. frederiksenii is a recently named organism that was previously a biogroup of *Y enterocolitica*. Its role in human infections is questionable, but it has rarely been isolated from blood, wound, and sputum cultures and must be considered a potential human pathogen. It has also been isolated from stool specimens, but its significance as an intestinal pathogen is yet to be established. The organism is commonly found in the environment. YY*frederiksenii is* characterized biochemically by positive tests for ornithine decarboxylase and fermentation of rhamnose and sucrose.

TABLE 10.8. BIOCHEMICAL AND SEROLOGICAL DIFFERENTIATION OF *SHIGELLA* SPECIES

	Reaction of			
Test	*S. dysenteriae*	*S. flexneri*	*S. boydii*	*S. sonnei*
Mannitol fermentation	—	+	+	+
ONPG	V	—	V	+
Ornithine decarboxylase	—	—	—	+
Serogroup	A	B	C	D

Y. kristensenii and *Y intermedia* are also new species that were formerly biogroups of *Y enterocolitica*. These organisms have been recovered from significant human sources, but their role in human infections has not been firmly established. *Y. kristensenii is* characterized by positive tests for ornithine decarboxylase and negative reactions for raffinose, rhamnose, and sucrose fermentation. *Y. intermedia is* ornithine, rhamnose, and sucrose positive. It is also positive for raffinose and melibiose fermentation at 25°C, and these reactions are helpful in distinguishing *Y. intermedia* from *Y. frederiksenii*.

Y. pestis is the well-known cause of the severe systemic illness, plague, which may be acquired directly or indirectly from infected animals (bubonic) or via aerosols from infected patients (pneumonic). Most infections in the United States are from accidental contact with an infected animal. *Y. pestis* is a dangerous organism, and extreme caution should be used in handling it in the laboratory. A controversy has existed over the nomenclature of this organism because of the recent proposal that *Y. pestis* should be considered a subspecies of *Y. pseudotuberculosis*. However, because of the extreme danger of serious infection from *Y. pestis* compared with the danger from YY *pseudotuberculosis,* clinical and public health laboratories should continue to report separate species. *Y. pestis is a* fastidious organism, and it is characterized by mostly negative biochemical tests. However, more than 90% of the strains ferment arabinose, mannitol, trehalose, and xylose. A negative urease test is important for distinguishing *Y. pestis* from *Y. pseudotuberculosis*.

SEROLOGICAL TESTING

Serological confirmation is an essential part of the identification of *Salmonella* and *Shigella* isolates, and it is the only means of recognizing certain pathogenic *E. coli* strains. Serological identification of these organisms is performed by slide agglutination techniques with polyvalent or specific somatic-antigen grouping sera followed, in some cases, by testing with individual sera for specific serotype identification. Serogroup testing should be performed by all clinical microbiology laboratories, at least for *Salmonella* and *Shigella* isolates. Serotype identification based on flagellar (H) antigen testing can be referred to public health laboratories in most instances. Because of the possibility of serological cross-reactions, it is essential that isolates be subjected to rigorous biochemical testing before serological analysis is performed.

Biochemical and serological testing are complementary, and both are required for identification of *Salmonella* and *Shigella* species and certain *E. coli* isolates.

Slide agglutination tests are done by making a dense suspension of the organism to be tested in a drop of saline (treated with phenol or mercuric iodine). Growth from a triple sugar iron slant is a convenient source for the preparation of the suspension. A drop of properly diluted antiserum is added to the organism suspension and mixed with a wooden stick. The slide is then rocked back and forth and observed for clumping of the suspension. A positive test is indicated by rapid, complete agglutination of the bacteria. Control suspensions without added antisera should also be tested to detect autoagglutination.

Salmonella Species

Isolates confirmed biochemically as members of the genus *Salmonella* should be tested with polyvalent antiserum. These reagents contain agglutinins for *Salmonella 0* antigens and Vi antigen, representing approximately 95% of the *Salmonella* isolates likely to be encountered in clinical laboratories. If agglutination occurs with the polyvalent reagent, testing with individual grouping reagents should be done. Isolates giving a positive test may be reported as a certain *Salmonella* serogroup and referred to a reference laboratory for serotyping. If an isolate is biochemically *a Salmonella* species but no agglutination is obtained with the polyvalent reagent, a suspension of the organism should be heated in a boiling-water bath for 15 min, cooled, and then retested. Heating removes capsular antigens that may block the reaction with the somatic antigen. In addition, up to 5% of strains will not agglutinate because they belong to a serogroup not represented in the reagents used. Such strains should be sent to a reference laboratory for further testing. Finally, the serological reactions of S. *typhi* are of special note. Unheated *S. typhi* suspensions will typically agglutinate in Vi antisera but not in D-grouping sera, whereas after heating, the reactions are reversed. Agglutination will occur in most polyvalent antisera with both heated and unheated suspensions because the polyvalent reagents contain agglutinins to both 0 group D and Vi antigens.

Shigella Species

Serological confirmation of *Shigella* isolates is based only on O-antigen testing and is therefore less complicated than *Salmonella* testing. *Shigella* isolates fall into four serogroups, A, B, C, and D, which correspond to the four recognized *Shigella* species, *S. dysenteriae, S. flexneri, S. boydii,* and *S. sonnei,* respectively. The O-group-ing sera for testing *Shigella* isolates are polyvalent (A, 10 of the 12 serotypes; B, 6 serotypes and two variants; and C, 15 of the 18 serotypes) except for the group-D reagent, which has a single serotype. Isolates biochemically identified as *Shigella* species should be tested by slide agglutination, as described above, for reactivity with the four grouping sera. Agglutination in one of the four sera plus the bioche-mical reactions of the isolate can be used for species identification, and results can be reported as *S. sonnei, S. flexneri,* etc. If the isolate is biochemically *a Shigella* species but fails to agglutinate in the grouping sera, it should be retested after boiling as described above. If heated suspensions fail to agglutinate and the isolate is well characterized biochemically, it should be sent to a reference laboratory for further testing. Particular care must be taken in the biochemical identification of isolates because of serological cross-reactions with *E. coli* and other *Entero bacteriaceae*.

E. Coil

Serotyping for the identification of *E. coli* strains capable of causing enteric disease has been applied mostly to the group of organisms known as "enteropathogenic." A limited number of the approximately 162 0 serogroups of *E. coli* have been associated with outbreaks of diarrhea in nurseries. However, these organisms have been infrequently recognized in recent years. Serotyping of *E. coli* strains is a complicated and difficult process and very seldom yields definitive etiological information. Accordingly, routine testing in hospital laboratories for the detection of enteropathogenic serotypes is no longer recommended. When nursery or other outbreaks of diarrheal disease occur and no specific pathogen can be identified, *E. coli* isolates may be sent to a reference laboratory for serotyping to possibly identify a cause of the out break. Specific serotypes have also been associated with enterotoxigenic and enteroinvasive strains of *E. coli,* but reagents have not yet become commercially available for routine testing in clinical laboratories.

Recently, a new enteric infection has been ascribed to *E. coli*. Hemorrhagic colitis is a distinctive form of severe, bloody diarrhea that often requires hospitalization. *E. coli* 0157:H7 strains have been implicated as the cause of this disease, and they can be definitively identified only by serotyping. Such serotyping is best performed at reference centers. Several representative *E. coli* colonies should be picked for testing, and a negative test for sorbitol fermentation may be useful in the selection of isolates for serotyping.

ENTEROTOXIN TESTING

Laboratory diagnosis of diarrhea due to *E. coli* poses a particular problem because of the ability of this organism to cause diarrhea by several different mechanisms. This problem is further complicated by the inability of traditional cultural and serological methods to detect many pathogenic strains. The detection of most pathogenic *E. coli* strains requires methods for toxin testing. Until recently, the available methods were not applicable to routine use in clinical laboratories because they required the use of animals or cell cultures. However, newly described methods can be used in clinical laboratories, especially if the reagents become commercially available. These methods are based on immunologic and DNA hybridization technologies.

Promising immunologic techniques include enzyme-linked immunosorbent assays and particle agglutination methods. Although a gel immunodiffusion technique is commercially available, it requires several days to perform and is technically demandin. Enzyme-linked immunosorbent assay methods involve the binding of toxin in culture filtrates to ganglioside-coated wells of plastic microtiter plates. The presence of the toxin is detected by the addition of specific antibody, followed by the addition of enzymeconjugated antibody. A positive test is indicated by a color reaction. Such tests have been found to have up to 100% sensitivity and specificity when compared with traditional cell culture assays. The second promising immunologic technique is coagglutination with protein A-bearing *Staphylococcus aureus* conjugated with antibody against *E. coli* enterotoxins. This technique has been found to be effective for the detection of enterotoxin in colony lysates of *E. coli*. Both techniques also have promise for the direct detection of enterotoxins in stool specimens without culture.

DNA hybridization has been successfully applied to the detection of *E. coli* enterotoxin from isolated colonies and directly in stool samples. The technique employs radioactively labeled DNA fragments that encode for the heat-stable and the heat-labile enterotoxins of *E. coli*. The application of these genetic probes to isolated colonies involves growing the organism to be tested on nitrocellulose filters, lysing the colonies, and hybridizing the colony DNA with the probe DNA. Hybridization is detected by autoradiography, and positive colonies are indicated by a black spot. For the direct detection of enterotoxins, stool specimens are spotted onto filter paper, incubated overnight, and tested as described above. These techniques have been very effective for the detection of both the heat-stable and the heat-labile toxins.

The availability of these methods makes enterotoxin testing feasible for many laboratories, and when these or similar methods become commercially available in the near future, it will be possible for all but the smallest laboratories to detect *E. coli* enterotoxins. Due to a lack until now of enterotoxin detection methods applicable to clinical laboratories, the importance of *E. coli* as a cause of diarrhea in the United States has been unclear. The availability of these newer methods should clarify the role of *E. coli* and lead to better diagnosis of bacterial diarrhea.

ANTIBIOTIC SUSCEPTIBILITY

The antibiotic susceptibility of the *Enterobacteriaceae is* variable according to the genus and species and, to a certain extent, within species. Accordingly, susceptibility testing should be done on individual isolates when antibiotic therapy is indicated. However, most genera and some species have susceptibility patterns that are predictable in general, and useful information may be gained from such patterns to guide antimicrobial therapy until susceptibility testing results are available. However, susceptibility patterns may vary among institutions, and profiles should be generated for each laboratory if possible. In addition, susceptibility patterns for individual species have been unstable in recent years. For example, until recently, *E. coli* was considered susceptible to most of the commonly tested drugs for gramnegative bacilli, but increasing numbers of isolates in recent years have been resistant to ampicillin, cephalothin, and other drugs. Therefore, it is important for laboratories to update their susceptibility profiles regularly.

The following general patterns of susceptibility hold true for most hospitals, although individual variation occurs as mentioned above. *Citrobacter diversus* is characteristically susceptible to all drugs commonly tested for gram-negative bacilli except ampicillin and carbenicillin. *Ci freundii* is more resistant overall, but twothirds of its isolates are susceptible to carbenicillin and more than one-third are susceptible to ampicillin. The most commonly encountered *Enterobacter* species, *E. aerogenes* and *E. cloacae,* are resistant to ampicillin and cephalothin. As mentioned above, most *E. coli* isolates are susceptible to the commonly tested drugs, but an increasing proportion are resistant, especially to ampicillin and cephalothin. In our experience, more than *40%* of *E. coli* isolates are resistant to ampicillin, and more than 20% are resistant to cephalothin. Such *E. coli* strains resemble *Enterobacter* species in their susceptibility profiles. *K pneumoniae* isolates are characteristically resistant to ampicillin and carbenicillin, but most isolates are susceptible to other drugs commonly tested. *K. oxytoca* isolates are more often susceptible to carbenicillin. *M. morganii* isolates are predictably resistant to ampicillin and cephalothin, and they are characteristically resistant to colistin. In addition, more than *40%* of the isolates are resistant to tetracycline. Isolates of *P. mirabilis* are among the most susceptible of the *Enterobacteriaceae,* but they are classically resistant to colistin and tetracycline. Isolates of *P. vulgaris* have a susceptibility pattern very similar to that of *M. morganii. Providencia* species tend to be relatively resistant to antimicrobial agents, and *Salmonella* and *Shigella* isolates in the United States are susceptible to most drugs, although multiply resistant strains are common in other countries. *S. marcescens,* especially hospital strains, is often drug resistant, and it is classically resistant to ampicillin, cephalothin, and colistin.

Some of the less commonly encountered species also have characteristic susceptibility patterns. *Cedecea* strains are usually colistin and cephalothin resistant. *K cryocrescens* is usually susceptible to carbenicillin and cephalothin, but *K. ascorbata* is generally resistant. *P. penneri* strains were initially recognized in part because of resistance to chloramphenicol, and *T. ptyseos* isolates are unique among the *Enterobacteriaceae* in their marked susceptibility to penicillin.

11

ADJUNCTIVE THERAPY

Adjunctive therapy for infections is very tempting. Its long history probably started with plant therapy and continued with the use of chemicals, purge, phlebotomy and blood transfusions in medieval medicine. Many specific agents were advertised in the past as a panacea only to have been discarded and forgotten.

Glucagon, heavy metal chelators, fibronectin, pentoxyphylline, antithrombin III, oxygen free radical scavengers, immune globulin, a,-antitripsin, surfactant, growth hormone, complement, nonsteroidal anti-inflammatory drugs (NSAID), and many other substances have recently been proposed as having beneficial effects. Therapies such as hemodynamic support with fluid resuscitation, inotropes and vasopressors, ventilatory support and nutrition are generally accepted. In this chapter we will discuss four adjunctive therapies: corticosteroids, immunotherapy including immunoglobulins and monoclonal antibodies, NSAID, and oxygen free radical scavengers. Most of the adjunctive therapies to be discussed have essentially been evaluated in adults.

CORTICOSTEROIDS

The rationale for the use of corticosteroids in sepsis is based on several properties of the drug. Corticosteroids stabilize cell membranes and lysosomal membranes. Pharmacological doses of steroids may also prevent the release of lysosomal enzymes involved in cell autolysis secondary to endotoxin. Glucocorticoids suppress inflammation by inhibition of leukocytic migration to the inflammatory site. These effects are believed to be beneficial as the host overresponds to the infectious insult. Some studies show that glucocorticoids inhibit complement activation by endotoxin, thereby preventing the deleterious effects of endotoxins, such as disseminated intravascular coagulopathy (DIC). Glucocorticoids through their effect on the arachidonic acid cascade, decrease pulmonary hypertension, permeability, and leukocyte agglutination, and prevent production of prostacyclin, but not thromboxane. Findings in animals support the proposal that glucocorticoids maintain hepatic gluconeogenesis during endotoxemia.

On the other hand, the use of corticosteroids carries with it important complications such as secondary infections due to immune suppression, psychosis, gastrointestinal bleeding, and hyperglycemia.

Numerous investigators documented increased survival in various animal models of gram-negative sepsis after high-dose glucocorticoid therapy. These studies indicated that the effect was more pronounced if the drug was given before or soon after the exposure to the bacteria or endotoxin. The beneficial effects of corticosteroids in humans were typically recorded in anecdotal reports.

Early studies demonstrated either reduced overall mortality, or a trend for reversal of shock if corticosteroid therapy was started within four hours of the onset of shock.

Two very important prospective, randomized, double-blind, placebocontrolled studies were published in 1987. These studies attempted to treat patients within a short period of time after the diagnosis of sepsis. Bone et al. reported a study of high-dose methylprednisolone sodium succinate (30 mg/kg) given every 6 h for a total of 4 doses; therapy started within 2 h of the diagnosis of severe sepsis. Three hundred and eighty-two patients were enrolled. The authors found no 'significant difference in the prevention of septic shock (steroids 46 vs. placebo 37%), the reversal of shock (steroids 65 vs. placebo 73%), or overall 14-day mortality (steroids 34 vs. placebo 25%, $p < 0.06$). Among patients treated with methylprednisolone, there was no difference in the incidence of secondary infections, but there were significantly more deaths related to secondary infection in the steroid-treated patients (steroids 34 vs. placebo 7%, $p < 0.05$). In the subgroup of patients with elevated serum creatinine levels (> 2 mg/dl), mortality was significantly increased among those patients receiving corticosteroids. A trend towards an increased incidence of the adult respiratory distress syndrome (ARDS) was seen in the corticosteroid-treated patients (32%) compared to the placebo group (25%). Significantly less steroid-treated patients (31%) reversed their ARDS compared to placebo-treated patients (61 %, $p < 0.01$), and the mortality of ARDS patients treated with steroids was significantly higher (52%) compared to placebo-treated ARDS patients (22%, $p<0.01$). Contrary to Sprung's study cited above, there was no indication that the early initiation of steroid therapy had any effect on shock reversal. This may be related to the fact that all patients were treated early.

The Veterans Administration Systemic Sepsis Cooperative Study group conducted a trial of early, high-dose methylprednisolone sodium succinate (30 mg/kg followed by an infusion of 5 mg/kg/h for 9 h) in 223 patients with clinical signs of systemic sepsis and a normal sensorium. The mortality rate was similar in the placebo (22%) and glucocorticoid groups (21 %). The resolution of secondary infections was faster in patients receiving placebo (52%) than in those receiving steroids (19%, $p<0.05$), but mortality rates were similar in both treatment groups for those patients with unresolved infections.

Some evidence suggests that corticosteroids may still be beneficial in specific diseases such as infants and children with bacterial meningitis and patients with severe typhoid fever. Hoffman et al., in a randomized, placebo-controlled, double-blind study of 38 patients, demonstrated that dexamethasone (3 mg/kg followed by a 1 mg/kg infusion every 6 h) reduced mortality from 56 to 10% ($p = 0.03$) in patients with culturepositive severe typhoid fever. As a result dexamethasone was recommended for patients suspected of typhoid fever with altered mental status or in shock.

A recently published placebo-controlled, double-blind study by Odio et al. in 101 infants supports the experimental and clinical evidence that dexamethasone reduces meningial inflammation and improves the outcome of patients with bacterial meningitis. The infants included in the study had culture-proven bacterial meningitis or clinical signs of meningitis and findings characteristic of bacterial infection on examination of the cerebrospinal fluid (CSF). The patients were randomly assigned to receive either dexamethasone and cefotaxime ($n = 52$) or cefotaxime plus placebo ($n = 49$). Dexamethasone (0.15 mg/kg) was given 15-20 min before the first dose of cefotaxime and was continued every 6 h thereafter for 4 days. The authors noted a marked clinical and laboratory improvement in the dexamethas-onetreated group after 12 and 24 h as compared to the placebo group.

The relative risk of sequelae for a child receiving placebo as compared with a child receiving dexamethasone was 3.8.

Based on the above data, except for bacterial meningitis in children and infants and severe typhoid fever, the medical consensus today is that the treatment of patients with sepsis, whether early or late (septic shock), should not include corticosteroid therapy. Corticosteroids do not prevent septic shock or improve survival in sepsis and are associated with increased morbidity and perhaps even mortality in certain subgroups of patients.

If one wants to use corticosteroids for adrenal insufficiency, which in fact is uncommon in sepsis, the requirement is approximately 300 mg of hydrocortisone a day and not the pharmacological doses of steroids that were given in the past.

IMMUNOTHERAPY

Immunoglobulins

Immunoglobulins (Ig) are an important component of host defenses against infections. These proteins are produced by B lymphocytes and plasma cells. They react as antibodies with specific antigens. There are five classes of Ig: G, A, M, D, and E. IgG is the major Ig in the blood involved in secondary responses to previously encountered antigens. The first immunoglobulin (IgG) preparations from human blood were used in 1952 to treat immune deficiency conditions. The first Ig preparations required intramuscular administration and effective blood levels were difficult to achieve. In 1981, the first Ig for intravenous use was introduced. Intravenous Ig (IVIG) has the following advantages compared to the intramuscular preparations: less painful injections, no sterile abscesses, no volume limitation, and higher blood levels. On the other hand, IVIG is more expensive, and requires venous access. The incidence of side effects with IVIG is usually less than 5%.

Various immunodeficiency states have been described as important in pediatric practice. Transient hypogammaglobulinemia normally occurs midway through the first year of life with a drop in IgG levels to as low as 350 mg/dl after maternal antibodies decline and after a delayed onset of IgG production. This drop in IgG levels can be even more pronounced in preterm infants as most maternal Ig cross the placenta during the last weeks of pregnancy.

The NIH Consensus Development Conference recommended IVIG for primary immunodeficiency (e.g. X-linked agammaglobulinemia),

200-800 mg/kg/month tailored to the individual patient in order to maintain a trough level in the blood of 500 mg/dl. For chronic lymphocytic leukemia, the maintenance of a level of 640 mg/dl was recommended. After bone marrow transplantation the dose of 500 mg/kg/week was found to be effective in reducing the infection rate, CMV pneumonia, and mortality, but not in preventing graft-versus-host disease in children as opposed to adults. The prophylactic use of IVIG (400 mg/kg) in symptomatic HIV-infected children under 13 years of age without hemophilia, with clinical or immunologic evidence of HIV, was found to increase the time free from serious bacterial infection in those patients with CD4+ lymphocyte counts of more than 0.2 x 10^9.

In a recent multicernter, double-blind study 500 mg/kg/day of IVIG was also found safe and effective in reducing the risk of nosocomial infection in a group of neonates weighing 500-1,750 g at birth. There was an average decrease of 6 days in the length of hospital stay for these premature infants treated with immunoglobulin.

A controlled randomized study showed that premature infants with group B streptococcal perinatal infection receiving IVIG and; itibiotics had a lower mortality rate than infants treated with antibiotics alone. IVIG may also be of value in treating overwhelming infections due to *Hemophilus influenzae, Streptococcus pneumoniae,* and *Niesseria meningitidis.* The cost effectiveness of IVIG in these and other infectious conditions remains to be established.

Present day preparations of IVIG are safe and most adverse reactions are mild, preventable, and easily treated. Side effects include: pyrogenic reactions, headache, myalgia, fever, chills, vomiting, cardiovascular reactions, hypersensitivity and anaphylactic reactions. Patients with primary antibody deficiency and patients with compromised cardiac function may be at increased risk for reactions. The occurrence of these side effects may be related to the volume and rate of the infusion. For patients with repeated severe reactions, hydrocortisone 1-2 mg/kg i.v. can be given 30 min prior to the IVIG infusion. At the present time, IVIG does not transmit hepatitis or HIV.

Another type of immune globulin available is hyperimmune globulin which is especially enriched with a particular antibody. The globulin is obtained by collecting plasma from either preselected donors or immunized volunteers. These preparations have been found to be effective in high-risk populations but require further evaluation before their use can be recommended in serious neonatal infections.

Monoclonal Antibodies

The poor outcome of patients with gram-negative sepsis has been attributed to the lipopolysaccharide (LPS or endotoxin) component of the outer cell wall of the bacteria. Endotoxin triggers many of the systemic reactions encountered in sepsis. Part of the LPS, called lipid A, is shared by all gram-negative bacilli. It has been hypothesized that antibodies against lipid A might protect against the toxic effects of gram-negative sepsis.

Monoclonal antibodies were first produced by fusion with a mouse myeloma cell line and as a result contain foreign (murine) proteins. These antibodies have been given to patients but have the disadvantage that repeated administration might result in the formation of murine antibodies which might interact with the monoclonal antibody. The first human antiserum was developed by immunizing volunteers with heat-inactivated cells of the J5 mutant of *Escherichia coli* 0111:B, which induced an immune response to the core polysaccharide-lipid A complex. These clinical trials were rather promising but the polyclonal J5 antiserum did not become commercially available because of mild toxicity to donors, ability of donors to donate only once and primarily because of the potential for transmitting infection with pooled human blood. The introduction of human monoclonal antibodies overcame many of these problems, allowing the production of large quantities of antibodies of known isotype and epitope specificity without the fear of transmitting infectious agents.

The human monoclonal IgM antibody HA-1A binds to the lipid A domain of endotoxin and has been shown to bind specifically to many endotoxins and to a broad range of clinical isolates of gram-negative bacteria. Ziegler et al. performed a prospective, multicenter, randomized, double-blind, placebo-controlled clinical trial of HA- 1A in patients with a presumptive diagnosis of gram-negative sepsis. In the group of 200 patients with gram-negative bacteremia, HA-1 A decreased mortality from 49% (45/92 placebo patients) to 30% (32/105 HA-1 A patients, p = 0.01). In the group with gram-negative bacteremia and shock at entry, HA- 1A decreased mortality from 57% (27/47 placebo patients) to 33% (10/54 HA-1 A patients, p = 0.017). HA- 1A treatment was also associated with a more rapid resolution of the major complications of sepsis, and with a higher incidence of patients being discharged from the hospital. No benefit of treatment with HA-IA was demonstrated in the total group of 543 patients or the 343 patients with sepsis who did not prove to have gram-negative bacteremia.

Greenman et al. assessed, in a prospective, multicenter, randomized double-blind, placebo-controlled clinical trial, a murine monoclonal IgM antibody (E5) to endotoxin in the treatment of gram-negative sepsis. Patients received an intravenous infusion of either a murine monoclonal antibody (2 mg/kg) or placebo, with a second dose given after 24 h. All patients received appropriate antibiotic therapy. Among the 316 patients with documented gram-negative sepsis, mortality at 30 days was not significantly different between treatment groups (placebo 41%, E5 antibody 38%). E5 treatment showed significantly greater survival among patients with gram-negative sepsis who were not in shock at study entry (relative risk 2.3, p = 0.01). These patients also had a greater resolution of individual organ failure, 54% (19/35) in the E5 group compared to 30% (8/27) in the placebo group (p = 0.05). E5 was also found to be safe.

Antibodies to Cytokines

Many of the toxic effects of bacterial LPS (endotoxin) are mediated by endogenous cytokines released by activated macrophages. These cytokines include tumor necrosis factor (TNF) which in turn induces the production of other mediators. The cytokines may act synergistically in producing some of the lethal effects of gram-negative sepsis.

Circulating levels of TNF were found to correlate with the severity of systemic reactions to endotoxin. Several animal studies demonstrated that antibodies to TNF may confer immunity against the deleterious effects of endotoxin. Exley et al. published positive results of a phase I study evaluating the safety of an IgG monoclonal antibody to recombinant human TNF in 14 patients with severe septic shock. The overall survival was not different from that predicted in patients of this kind. Multicenter studies examining the efficacy of monoclonal antibodies to TNF are now in progress.

Ohlsson et al. believe that IL-1 plays a crucial role in endotoxic shock and that IL-i receptor antagonists may reduce mortality in severe sepsis. Preliminary trials of IL-1 receptor antagonist in patients with sepsis are also in progress.

Another mediator of sepsis is IL-6. This glycoprotein, derived from T cells, induces production of acute-phase proteins. Antibodies to IL-6 have been studied in animal models with promising results.

NONSTEROIDAL ANTI-INFLAMMATORY DRUGS

NSAID are believed to confer improved hemodynamics in endotoxic shock and to improve survival. The effect of NSAID has been extensively studied in animals. The mechanisms by which they produce these effects include inhibition of cyclooxygenase and phosphodiesterase activity. NSAID inhibit cyclooxygenase, thereby preventing the production of the end products of the arachidonic acid cascade - thromboxane and prostacyclin. Free radical superoxide is formed during this cascade by cyclooxygenase, and is later converted to hydroxyl free radicals. These oxidizing agents damage the membranes during the inflammatory process. The inhibition of phosphodiesterase increases the intracellular concentration of cyclic AMP, a substance which stabilizes the membranes. The potential hazard of using NSAID in sepsis is their ability to alter renal perfusion and decrease renal function through prostaglandin inhibition. The role of NSAID in the pathogenesis of peptic disorders is also well known.

Few studies have been performed to evaluate the effect of NSAID in humans in sepsis or septic shock. Michie et al. examined the effect of pretreatment with ibuprofen in 13 healthy men after intravenous administration of E. *coli* endotoxin. Ibuprofen did not prevent the rise in circulating TNF but greatly attenuated fever, tachycardia, and the production of stress hormones after endotoxin administration. Bernard et al. conducted a randomized multicenter trial of short course ibuprofen (three doses of 800 mg, every 4 h administered as a rectal solution). Signs of shock reversal including improved blood pressure, decreased heart rate, decreased minute ventilation, and decreased peak airway pressure occurred within 48 h in 43% of placebo patients and 88% of ibuprofen patients who had shock at entry. Oxygenation was not improved by the treatment, nor did the patients develop signs of renal toxicity.

OXYGEN FREE RADICALS

Oxygen free radicals produced during infection originate from neutrophils, and electron transport systems of other cells. Some of the important pathophysiological changes in sepsis are attributed to the production of oxygen free radicals. Oxygen free radicals damage the cell membrane by lipid peroxidation. Several new oxygen free radical scavengers, known as lazaroids, have recently been introduced. These substances are based on a 21aminosteroid structure, without glucocorticoid activity. In animal models, oxygen free radical scavengers are reported to be of value in sepsis by improving microcirculatory flow and increasing survival. Human clinical trials are expected in the near future.

12

IMMUNIZATIONS

Vaccines represent one of the most effective measures in preventive medicine. In recent years significant strides in the prevention of childhood illnesses have been made as the result of new and improved vaccines. In many countries children enjoy better health because of effective immunization programs which have diminished the morbidity and mortality of oncecommon transmissible diseases. To achieve this progress in child health, scientific technology and medical practice have combined efforts to understand the biology of infectious agents; to purify the agents and, in some cases, their components; to develop and test vaccines; to manufacture and administer these vaccines to appropriate segments of the population; to develop appropriate recommendations for immunization; and to implement schedules for immunization.

DISEASES FOR WHICH NEW VACCINES ARE AVAILABLE

Haemophilus influenzae, *Type b*

Haemophilus influenzae, type b (HIB) is a major cause of invasive bacterial infections in young children. It is the most common cause of bacterial meningitis and epiglottitis, and an important cause of septic arthritis, occult febrile bacteremia, cellulitis and pneumonia in children less than 5 years old, causing an estimated 11,000 cases of meningitis and 5,000 additional cases of invasive HIB disease annually in the

United States. The peak incidence of HIB meningitis in this country occurs between 6 and 12 months of age. In high-risk populations, such as Native Americans and children in New Guinea, rates of disease are higher and a greater proportion of cases of meningitis occur in the first year of life.

Because the majority of cases of HIB disease occurs in infancy, vaccines that provide protection by 6 months of age or earlier are necessary for effective control of HIB disease. Realization of this goal has been made possible by the development of conjugate vaccines in which the capsular polysaccharide of HIB, called polyribosylribitol phosphate (PRP), the major virulence factor and the antigen against which protective antibodies are directed, is chemically linked to a protein carrier.

The first HIB vaccine licensed was purified PRP and was recommended for administration to children 18-24 months of age or older. This nonconjugate vaccine was poorly immunogenic in nearly all children younger than 18 months of age and did not reliably induce protection in most children until 2 years of age or older. In a large Finnish trial of this vaccine, the protective efficacy was more than 90% in children vaccinated between 24 and 71 months of age and was ineffective in children less than 18 months of age. Serological findings in this study indicated that a serum anti-PRP concentration of > 1.0 ug/ml 3 weeks postvaccination was protective.

The differences in immunological properties between the purified pol} saccharide and protein-conjugated vaccines explain the inherent limitation of the first-generation HIB vaccine of PRP and the potential of the conjugal vaccines for the prevention of HIB disease. Purified PRP (and other polysa(charide antigens) has properties characteristic of a T-cell-independent ant: gen, as humoral responses appear to be mediated by B lymphocytes alon without T-helper lymphocytes. T-cell-independent antigens are generall poor immunogens in children less than 2 years of age. In contrast, th polysaccharide-protein conjugate vaccines are T-cell-dependent antigens i which T-helper lymphocyte activation as well as B-cell mediation of th humoral antibody response occurs. T-cell-dependent antigens elicit antibod responses in infancy and also booster responses, which are important for th development of effective polysaccharide vaccines.

Conjugate HIB vaccines produce protective antibody in infants and young children, and significantly greater concentrations of circulating anti-PRP at all ages than does unconjugated PRP and have

TABLE 12.1. *H. INFLUENZAE* TYPE B CONJUGATE VACCINES

Vaccine manufacturer	*Abbreviation for vaccine (trade name)*	*Carrier protein*	*Polysaccharide size*	*Spacer*	*Licensure in the United States (as of Dec, 1991)*
Connaught	PRP-D (ProHIBit™)	Diphtheria toxoid	medium	6 carbon	for children ?15 months of age
Praxis/Lederle	HbOC (HibTITER™)	CRM_{197} (a nontoxic mutant diphtheria toxin)	small (oligo)	none	for infants and children >2 months of age
Merck Sharp & Dohme	PRP-OMP .(PedvaxHIB™)	an outer membrane protein (OMP) complex of *N. meningitidis*	large	complex, involving a thioether	for infants and children >2 months of age
Pasteur Merieux	PRP-T	tetanus toxoid	large	6 carbon	none

replaced the purified PRP vaccines. For the conjugate vaccines, the assumption that a serum anti-PRP concentration of > 1.0 μg/ml is a serological surrogate of protection from immunization with conjugate vaccines has not been proven, and lower concentrations of anti-PRP following immunization may confer protection.

Children 15 months of age or older respond well to a single dose of any of the four conjugate vaccines. Each conjugate is comprised of a PRP antigen conjugated to a protein carrier. The different protein carrier, size of the saccharide component, and chemical linkage of each vaccine is summarized in the table. Of the four conjugate vaccines that have been developed, three have been licensed in the United States by the end of 1991.

In late 1990, two conjugate vaccines, an oligosaccharide-CRM_{197} (a nontoxic mutant diphtheria toxin) conjugate vaccine (HbOC) and an outer membrane *N. meningitidis* protein conjugate vaccine (PRP-OMP), were approved in the United States for use in infants beginning at 2 months of age on the basis of placebo-controlled field trials which demonstrated nearly 100% protection. In a study in northern California of HbOC, vaccine efficacy was 100% for infants receiving the 3-dose schedule at 2, 4 and 6 months of age. In a Navajo population of infants at high risk of HIB disease who were vaccinated at 2 and 4 months of age with either PRP-OMP or placebo, vaccine efficacy was 100% at 1 year of age and 93% in total.

A third conjugate vaccine, diphtheria toxoid conjugate vaccine (PRP-D), is licensed in this country for use only in children 15 months of age or older. It is not currently used in infancy because of conflicting results in clinical trials in young children. In Finland in a randomized, controlled trial in 114,000 infants who were vaccinated at 3, 4 and 6 months of age, efficacy was 94%, whereas in Ałaskan Native infants vaccinated at 2, 4 and 6 months, efficacy was 35%. The 95% confidence limits of -57% to 73% in the latter trial indicate lack of the efficacy of PRP-D in this population.

Routine vaccination against HIB disease with either of the two conjugate vaccines approved for use in infants, HbOC or PRP-OMP, is recommended for all children in the United States beginning at approximately 2 months of age. These conjugate vaccines have different properties leading to the different recommended schedules of administration. PRP-OMP gives good antibody response after the first dose. Following a single dose of PRP-OMP in infants 2-6 months of age, 78% achieve an antibody concentration of > 1.0 tg/ml. Only two doses are recommended in the first year of life, at 2 and 4 months, with a booster dose at 12 months. In prelicensure studies, HbOC elicited serum antibody concentrations of > 1.0.tg/ml after two doses in more than 90% of recipients of a 3-dose schedule beginning at 1-6 months of age; after three doses more than 98% have serum anti-PRP concentrations of ? 1.0.tg/ml. HbOC is recommended at 2, 4 and 6 months of age with an additional dose at 15 months. The 15-month dose is a reinforcing dose and ensures immunity. It may be given as any one of the conjugate vaccines, since all are immunogenic at that age. Because the different conjugate vaccines generate different immune responses and data on interchangeability are not available, infants under 15 months of age should complete the schedule of vaccination with the vaccine which was used to begin the immunization series. For children not immunized in the first 6 months of life, recommended schedules for primary immunization vary according to the child's age and conjugate vaccine to be given. According to United States recommendations, children who have not been immunized by 15 months of age should be immunized with a single dose of any licensed conjugate *Haemophilus* vaccine.

Pertussis

Pertussis continues to cause significant morbidity and mortality worldwide among young children. More than 60 million cases of pertussis and more than half a million deaths are estimated to occur yearly. In the United States, the number of cases has been reduced by more than 98% from the prevaccine era. Nevertheless, approximately 3,000-4,000 reported cases of pertussis still occur each year.

The experiences of countries where rates of pertussis vaccination have markedly declined provides strong support for continued routine immunization of infants and young children. For example, in the United Kingdom as a result of adverse publicity about pertussis vaccination, a decrease in immunization rates in 2-year-old children from 77% in 1974 to 30% in 1978 was followed in 1977-1979 by an epidemic of 102,500 cases of pertussis and in 1982-1983 by a second epidemic. A similar experience occurred in Japân where 13,105 cases and 41 deaths occurred in 1979 after routine immunization had been temporarily suspended in 1975.

Effective primary preventive programs necessitate immunization of young infants because the morbidity and mortality of pertussis is greatest in infants, especially for those less than 6 months of age.

Approximately 35% of reported cases in the United States occur in infants less than 6 months of age. In this age group, the case fatality rate is a44ioximately 1%, T ate hospitalized and complications such as pneumonia (17%), seizures (2.6%), and encephalopathy (1.1 %) are frequent. High rates of immunization in children beyond infancy may further reduce the risk of infection in infants by decreasing the incidence of infection in older family members and resulting transmission of *Bordetella pertussis* within the household.

The pertussis vaccine currently available in the United States is a suspension of inactivated *Bordetella pertussis (i.e.* whole cell vaccine) combined with diphtheria and tetanus toxoids. Studies of household contacts exposed to pertussis indicate that the efficacy of whole cell vaccine is 80% or greater. Serological correlates between immunogenicity and efficacy have not been established. Reactions to the whole cell vaccine are common. Local and febrile reactions occur in more than half of the DTP recipients. More serious reactions are much less common. They include prolonged crying, reported to occur in 3% of DTP recipients in a large study, high fever >40.5 °C (0.3%), hypotonic-hyporesponsive state (1:1,750 immunizations) and seizures (1:1,750 immunizations). These convulsions are usually associated with fever and reflect preexisting, underlying febrile or other seizure disorders.

Pertussis vaccine has evoked public controversy in several countries because of the temporally associated neurologic events and deaths after pertussis immunization. To determine the possible role of pertussis vaccine in permanent neurologic injury the British conducted the National Childhood Encephalopathy Study (NCES) from 1976 to 1979 to assess children who presented with serious neurologic disease and potential brain damage temporally associated with DTP vaccination. Thirty children had received DTP within 7 days of onset of their disorder. These children were agematched with controls. The data was interpreted by the authors to indicate a statistically significant risk from DTP vaccination of subsequent onset of chronic neurologic disease, estimated to be one in every 330,000 doses of DTP. These findings have been questioned and reexamined by a committee convened by the Institute *of* Medicine *of* the United States National Academy of Sciences. It concluded that a causal relationship between DTP vaccine and permanent brain damage had not been demonstrated. Children originally thought to have brain damage were either normal on further follow-up or had an alternative explanation, such as Reye syndrome or viral infection. Four other studies in the United Kingdom and the United States have also failed to provide evidence of a causal relationship between DTP vaccination and either serious acute neurologic illness or permanent neurological sequelae. The Institute of Medicine as well as the United States Centers for Disease Control, the American Academy of Pediatrics and vaccine advisory boards in the United Kingdom and Canada have concluded that although DTP may rarely produce neurologic symptoms, a causal relationship between DTP vaccine and permanent brain damage has not been demonstrated.

The side effects of whole cell pertussis vaccine prompted a search for pertussis vaccines with less reactogenicity. Acellular pertussis vaccines consisting of components of the pertussis organism rather than the whole bacteria have been developed and are presently in use in Japan. Two types of acellular vaccines have been produced; the first, called the B type vaccine, contains lymphocytosis-promoting factor (LPF) and filamentous hemagglutinin (FHA) and the second, or T type vaccine, contains LPF, FHA and agglutinogens. The proteins which make up the acellular vaccines are purified from culture supernatants of *Bordetella pertussis*. The antigen content and immunogenicity of the different vaccines vary; the B type vaccines elicit higher antibody responses to LPF than T type vaccines, both B and T type acellular vaccines induce higher FHA responses than whole cell vaccines and antibody to agglutinogens being induced only by T type vaccines. At present no data suggests that the two types of vaccine differ significantly in side effects or efficacy.

The Japanese have routinely used acellular pertussis vaccine in 2-year-old children since 1981 and have found the vaccine to be safe. Its use has been associated with a decline in reported cases of pertussis. Acellular pertussis vaccines have undergone clinical trials in Japan, Sweden and the United States and have been demonstrated to produce less fever and local reactions than whole cell vaccine. Assessment of the efficacy of pertussis vaccines is complicated by differences in the case definition of pertussis used in these studies. In a retrospective study of attack rates in Japanese household contacts, the efficacy of acellular pertussis vaccine in 24-month-olds was calculated to be 80-90%. A large-scale Swedish trial in children receiving 2 doses of acellular pertussis vaccine between 5 and 11 months of age found vaccine efficacy against pertussis ranging from 69 to 95% depending on the definition of pertussis used. A disappointment in the Swedish trial was the failure to correlate concentrations of circulating anti-LPF or anti-FHA with protection against pertussis after immunization.

No direct comparisons of efficacy with whole cell pertussis vaccine have been performed.

Questions remain about the use of acellular pertussis vaccines. First, which vaccine should be used? At present at least six acellular vaccines are produced in Japan which vary in antigen content. The critical antigens to include in a vaccine have not been identified since the antibodies correlating, with protection are not known. Second, will administration to infants at 2 months of age be safe and effective? At present little information on the safety or efficacy of acellular pertussis vaccine in infants less than 6 months of age is available but a large efficacy trial sponsored by the US National Institutes of Health is scheduled to begin in 1992 to assess the safety and efficacy of several acellular vaccines in this age group. Until the results of this trial are available, one or more acellular pertussis vaccines will be licensed in the United States in 1992 only for the doses at 15-18 months and 4-6 years of age. In addition, Japanese authorities have recently recommended initiation of pertussis vaccination with acellular vaccine in early infancy.

Polio

Poliomyelitis, an acute viral infection of the central nervous system that involves the spinal motor neurons and may result in transient or permanent paralysis of one or more extremities, is usually caused by one of the three serotypes of polioviruses. In the United States and other developed countries poliomyelitis due to naturally occurring (wild type) polioviruses has been almost eliminated by widespread immunization, but rare cases attributed to the live attenuated oral poliovirus vaccine are a continuing concern.

Oral polio vaccine (OPV) was licensed for routine use in the United States in 1962. OPV induces serum neutralizing antibody in 86-100% of infants after 2 doses of vaccine. These neutralizing antibodies protect against neurotropic poliovirus disease and prevent or reduce oropharyngea shedding of poliovirus on reinfection. Neutralizing antibody, however limits fecal shedding of virus on reinfection only when present in high tite. OPV also induces poliovirus-specific IgA in nasopharyngeal and intes tinal secretions. This IgA inhibits virus shedding from the gastrointestina tract during reinfection but this immunity may wane over time. OPV, after or more doses, has been demonstrated to have a 98% efficacy in preventinj poliomyelitis.

Inactivated polio vaccine (IPV) was originally developed in the

1950s However, within the past decade advances in cell culture technique have enabled the production of higher potency IPV. Since the serologic response to IPV is directly related to the antigen content of the vaccine, 'enhanced. potency' vaccines (E-IPVs) are more immunogenic than earlier IPVs and produce seroconversion rates equal to that of OPV. E-IPV differs from OPV in that no local IgA is produced in the gastrointestinal tract and poliovirus i readily shed in the feces following reinfection. E-IPVs were first introduced in the early 1980s in the Netherlands, France and Canada. Efficacy of E-IPV for preventing poliomyelitis is estimated to be 95%.

In the United States the last known case due to indigenous wild viru! occurred in 1979 and no evidence of continuing endemic transmission of wild type. poliovirus exists indicating that naturally occurring·poliovirus infection has been eradicated. However, an average of 5-8 cases of poliomy elitis caused by OPV virus occur annually. Only 35% of OPV-associates poliomyelitis cases are among recent OPV recipients, the majority of whicl occur following the first dose of vaccine. Approximately 50% of cases occw among close contacts of OPV recipients. About 10-15% of cases occur among OPV recipients or contacts who are immunodeficient and 5% of case occur in persons with no direct contact with OPV. The risk of paralysis i estimated to be one case in 520,000 first doses distributed and one case pe: 12.3 million subsequent OPV doses.

The continued annual occurrence of paralytic disease among OP\ recipients has perpetuated the controversy in the United States over it, policy of exclusive use of OPV and led to consideration of a change it vaccine strategies. Those favoring OPV cite the following advantages: (1) in duction of local (intestinal) immunity; (2) simple and painless administra tion; (3) secondary immunization of some contacts by fecal-oral spread thereby contributing to herd immunity; (4) ease of production, and (5) rela tive low cost. Proponents of IPV argue that IPV does not cause vaccine associated disease and exclusive use of IPV reduced the incidence of paralytic polio by more than 90% in the United States prior to the introduction of OPV and achieved eradication of indigenous polio in other countries, including Sweden, Finland, and the Netherlands. However, the exclusive use of IPV may enhance the likelihood of introduction and spread of wild type polioviruses in the community with increased risk of disease in the unimmunized as has occurred in countries using IPV, such as Finland and Israel.

A third strategy would take advantage of the benefits of both IPV and OPV by a schedule of sequential administration of IPV and OPV. Studies have shown that recipients of E-IPV followed by OPV have serum neutralizing antibodies that are equal or superior to E-IPV or OPV alone. Reevaluation of polio vaccine policy in the United States was undertaken in 1988 by a panel appointed by the Institute of Medicine of the National Academy of Sciences. The panel concluded that consideration should be given to a vaccine schedule in which two doses of E-IPV would be given initially in order to prevent vaccine-associated paralysis, followed by OPV at 18 months and at school entry to provide the pharyngeal and intestinal immunity necessary to reduce circulation of the wild virus in the population and allow sufficient excretion of vaccine virus to immunize susceptible unimmunized contacts. The panel concluded such a sequential schedule should be instituted when IPV could be effectively combined with a DTP preparation. Further studies of the immune response and viral shedding following OPV in recipients of E-IPV and OPV are in progress to evaluate this strategy.

Polioviruses continue to circulate and cause paralysis in the developing world, especially those countries in the tropics. Most cases of paralytic polio in developing countries occur in children between the ages of 6 months and 2 years. OPV is used almost exclusively in the developing world because of lower cost and ease of administration and because the potential transmission to unimmunized contacts may boost immunity levels in the population. In 1974 the World Health Organization recommended that all developing nations institute a 3-dose OPV schedule in the first 7 months of life. In 1985, the schedule was amended to recommend that the first dose be given at birth. Coverage has been less than ideal but the concept of 'National Vaccination Days' has improved immunization rates in many countries. The efficacy of OPV in developing countries has been variable. The basis for the less than expected efficacy rates is poorly understood but may relate to lower potency OPV than that used in the US.

The potential use of E-IPV in developing countries is undergoing examination. A study in Senegal found lower than expected vaccine efficacy with rates of 36% after one dose and 89% after two doses despite adequate seroconversion. The use of E-IPV combined with OPV reduced the incidence of paralytic disease in Israel where an epidemic of polioviru type 1 in the Gaza Strip occurred despite high rates of OPV vaccination The problem of reduced efficacy of OPV in the developing world may bi solved by giving E-IPV simultaneously with OPV, especially if E-IPV coup be combined with DTP.

Typhoid Fever

The public health importance of typhoid fever is substantial in the developing world. Approximately 33 million cases per year occur worldwid4. In developing countries mass immunization is potentially effective it control of typhoid fever.

Parenteral inactivated vaccines have been used for many years. In fiek trials of inactivated typhoid vaccine, vaccine efficacy has ranged from 51 to 76%. Protection has not been correlated with specific *Salmonella typhi* antibodies. In addition, protection in experimental challenge studies in volunteers has varied with the challenge inoculum and can be overcome by a high inoculum of *S. typhi*. Following parenteral vaccination, febrile reactions, headache and severe local pain and swelling are common. Of vaccinees, 13-24% have subsequently missed school or work. Booster doses are recommended for recipients of parenteral vaccines.

Two new typhoid vaccines that provide significant protection withou causing adverse reactions have been licensed in many countries. These vaccines include an attenuated *S. typhi* strain, Ty 21 a, a live oral vaccine, and the purified Vi polysaccharide capsular antigen of *S. typhi,* given parenterally.

The oral vaccine consists of a stable mutant of a pathogenic *S. typhi*, Ty 21 a, which lacks the enzyme UDP-galactose-4-epimerase and the Vi capsular polysaccharide. Efficacy of the oral Ty 21 a vaccine in clinical trials has ranged from 42 to 96% following the initial series of three doses with the lower efficacies seen in trials from areas with highly endemic disease. The mechanism of protection is not known. Reactions from the oral Ty 21a vaccine have been mild and rare. In safety trials, adverse reactions occurred with equal frequency among groups receiving vaccine or placebo. Since data on safety and efficacy of young children is not available, the manufacturer currently recommends that Ty 21 a vaccine should not be given to children less than 6 years old. Recommendations for booster doses for those whose primary immunization was with oral vaccines have not yet been determined. Primary immunization involves multiple doses and varies in schedule and dose according to the vaccine preparation. Since the oral vaccine is a live-attenuated strain, it should not be given to immunocompromised persons, including those known to be infected with HIV. This vaccine is promising for the control of endemic typhoid fever because protection is for at least 5 years and appears to be prolonged, and since mass immunization may result in herd immunity.

Purified Vi, a recognized virulence factor of *S. typhi,* has been used as a parenteral polysaccharide vaccine and has been as well tolerated as other polysaccharide vaccines. Efficacy of the Vi vaccine in clinical trials is 72% at 17 months and 64% at 21 months following a single dose. Since inoculation with additional doses of purified Vi vaccine fail to boost antibody titers Robbins and co-workers have conjugated the Vi polysaccharide to a carrier protein to confer T-cell-dependent properties upon the antigen. Clinical trials with Vi protein conjugates are in progress.

DISEASES IN WHICH IMMUNIZATION STRATEGY HAS CHANGED

In developing recommendations for immunization, multiple factors are considered, including vaccine characteristics, immunological responses to vaccine antigens, assessment of the benefits of the vaccine, risk and epidemiology of the disease and its complications, vaccine costs, risks of adverse reactions and recommended ages of routine health visits.

Changes in relative benefits and risks necessitate continued evaluation of recommendations. As new vaccines and scientific knowledge become available, schedules are modified and changed. Examples include the introduction in 1992 of universal infant immunization for hepatitis B in the United States because of the failure of selective immunization to significantly decrease rates of hepatitis B and the recommendation of a second dose of measles vaccine in the United States in response to the marked upsurge in measles cases in the late 1980s.

Hepatitis B

In the United States an estimated 200,000 to 300,000 cases of hepatitis B (HBV) occur each year. The majority of cases occur in persons with risk factors, which include immigration from a country where hepatitis B is highly endemic, multiple transfusions, or a high-risk life-style involving intravenous drug use or contact with multiple sexual partners. However, approximately 30-40% of acute HBV infections occur in persons with no identifiable risk factors. The incidence increases rapidly in sexually active adolescents and young adults. Between 3.3 and 25% of all America have had a hepatitis B infection by ages 25-34.

The recommended strategy for control of hepatitis B virus infection in the United States was changed in late 1991 from selective to a universal immunization. The previous strategy for control of HBV recommended by the National Centers for Disease Control (CDC) was based on selective vaccination of high-risk populations and serologic screening of all pregnant women for HBsAg. However, since 1982 when vaccine became available, the reported incidence of HBV infection has not significantly declined and prompted review of the strategy by the CDC and the American Academy of Pediatrics.

Failure to control HBV infections in the United States is due to several factors. First, many patients are not aware of their risk and therefore do not identify themselves as a high-risk patient to their physician or health care facility. For, example, adolescents beginning sexual activity are often unaware of the risks of HBV involved in heterosexual activity with multiple partners or homosexual experiences. Second, adolescents and young adults, those at highest risk of HBV infection, do not regularly present for routine health care or immunizations. Third, many health care providers are often not aware of groups at high risk for HBV infection and frequently fail to identify potential candidates for vaccination during routine health care visits. Finally, no known exposure or risk factor can be identified for one-third to one-half of reported HBV cases.

The failure to control HBV infection in the United States led to the recommendation by the CDC in December 1991 for universal childhood immunization. The advantages of this approach include the existence of effective programs of routine childhood immunization, protection without the need to identify specific risk factors and protection before significant exposure occurs. The impact of this strategy will not occur for several decades and will need to be supplemented during this interval by continued vaccination of high-risk persons and, resources permitting, routine immunization of adolescents. The cost of a program of infant immunization is less than that for adolescents and adults, since lower doses are given than those for the older population. In addition, HBV vaccine may be combined with DTP and HIB making immunization easier and less costly. Such products are currently under development.

The recommended doses and schedule for the use of hepatitis B vaccines licensed in United States in infants and other age groups are shown in the table. Routine screening of pregnant women for HBsAg is still recommended because the dose and schedule of the vaccine given to the infant as well as the use of hepatitis B immune globulin (HBIG) are based on the HBsAg status of the mother. Older children and adolescents at increased risk of exposure to hepatitis B should be immunized. The universal immunization of all adolescents is also recommended if resources are available.

TABLE 12.2. RECOMMENDED DOSES AND SCHEDULES OF HEPATITIS B VIRUS VACCINES IN THE UNITED STATES

Group	*Dose of Recombivax HB Merck, Sharp & Dohme*		*Dose of Engerix-B SmithKline Beecham*	
	µg	**ml**	**µg**	**ml**
Infants of HBsAg-negative mothers	2.5	0.25	10	0.5
Infants of HBsAg-positive mothers	5	0.5	10	0.5
Children <I 1 years	2.5	0.25	10	0.5
Children and adolescents 11-19 years	5	0.5	20	1.0
Adults a:20 years	10	1.0	20	1.0
Dialysis patients and other immunocompromised persons	40	1.0[1]	40	2.0

Hepatitis B is a major health problem worldwide with more than 2 million new cases reported yearly. High endemic areas of the world include sub-Saharan Africa, southeast Asia and Oceania. The age of transmission varies in different countries. Perinatal transmission is most frequent in the Far East and early childhood acquisition of infection is more common in Africa. Chronic infection is most common when hepatitis B is acquired perinatally or in early childhood and hepatitis B, as a result, is a major cause of adult death in most of Africa and much of Asia by causing liver disease and hepatocellular carcinoma. To prevent HBV infection at least 20 countries now recommend HBV vaccination for all infants. HBV vaccination may be integrated into the Expanded Programme on Immunization (EPI) of the World Health Organization when HBV vaccine can be produced on a large scale at affordable prices.

Measles

In the 20 years after the introduction of measles vaccine in the United States in 1963, the reported incidence of measles decreased by 99%. However, measles in the United States underwent a resurgence in the late 1980s, leading to changes in the recommendations for measles vaccine in 1989.

Measles in the 1980s occurred in two patterns. The first was outbreaks in junior and senior high school students and in college students, most of whom had been vaccinated. They represented those persons not protected after a single dose of measles vaccine. Recent data indicate that the rate of primary vaccine failure is 2-5%. The second pattern was of outbreaks in preschoolers, primarily in inner cities, the majority of whom were unvaccinated. Many cases occurred in children younger than the routine age for immunization in this country. As a result of the problem of vaccine failure, the American Academy of Pediatrics and the CDC recommend a second dose of measles-mumps-rubella vaccine (MMR) for all children to immunize those who did not seroconvert after the first measles immunization. The success of such a 2-dose schedule in preventing measles has been demonstrated in several European countries.

The recommended age for the second dose of measles vaccine by the American Academy of Pediatrics is at entry to junior high or middle school, since most vaccine failures have occurred in older children and vaccination of older children will have the most rapid impact on disease occurrence. In addition, this recommended age is consistent with the goals of improved adolescent preventive health. In contrast, the CDC recommend the second dose of MMR to be given at school entry. The CDC recommendations are directed toward children vaccinated in public health clinics and were made for ease of implementation and enforcement. These children are already receiving DTP and polio at school entry, and school entry laws provide the mechanism for enforcement. The major disadvantage is the delay between implementation and impact on disease occurrence. While the recommended age by these two bodies differs, both acknowledge the advantages and validity of the other's recommendation.

Measles cases and epidemics continued to increase in 1989 and 1990 despite the implementation of the 2-dose recommendation because of the failure to vaccinate substantial numbers of young children, especially inner city populations of black and Hispanic children. In these groups the risk of measles was 7- to 9-fold higher than white children. Many of these cases are the result of missed opportunities for vaccination. The resulting strategy for control of measles is to improve delivery of vaccine, especially to children in the inner city. Physicians need to assure that children are vaccinated on time or are recalled or tracked if they have missed appointments. Children should receive all the vaccinations they need in the fewest possible visits and vaccines should not be withheld in children with minor illnesses such as mild diarrhea or an upper respiratory infection. In addition, barriers to vaccination such as requiring scheduled appointments or prior physical examinations should be removed.

Despite the availability of effective measles vaccines and successful immunization programs in many countries, measles is still the most common cause of death due to a vaccine-preventable disease among children of the world. WHO estimated that in 1989,49 million cases of measles and 1.5 million deaths due to measles and its complications occurred. The high morbidity and mortality of measles in developing countries among infants younger than 9 months, the usual recommended age of vaccination by WHO, has stimulated research to successfully immunize younger infants. Alternative measles vaccine strains with higher potency than the Moraten strain, the strain currently in use in the United States, have been evaluated. A high titer Edmondston-Zagreb strain was found to have seroconversion rates in infants at 4-6 months of age equivalent to that of other vaccine strains when given at 9 months of age. In developing countries where measles causes significant morbidity and mortality in infants less than 9 months of age, the Global Advisory Group of the WHO Expanded Programme on Immunization (EPI) now recommends hightiter Edmondston-Zagreb measles vaccine at 6 months of age.

DISEASES FOR WHICH IMPROVED VACCINES ARE UNDER DEVELOPMENT

Cholera

Cholera remains a significant public health concern in developing countries. The currently available vaccines are inactivated preparations of wholecell suspensions of classic Inaba and Ogawa strains of *Vibrio cholerae* and are used parenterally. In field trials conducted in cholera endemic regions, these vaccines have been shown to be only about 50% effective in reducing ti incidence of clinical illness for a period of 3-6 months after vaccination [Because of the lack of efficacy of parenterally administered vaccines choler vaccines that elicit gastrointestinal immune responses directed at choler antigens or toxin are now under development.

Inactivated oral vaccines, a killed whole-cell *V. cholerae* combined with cholera toxin B subunit and a killed whole *V. cholerae* 01, have been shown in field trials to have better efficacy than the parenterally administered killed vaccine presently in use. However, protection in young children, who are most likely to die from natural infection, is less than adequate, and multiple boosters must be administered.

Various preparations of live oral cholera vaccines are under development and include live attenuated *V. cholerae* O1 vaccines and a typhoid-cholera hybrid vaccine derived from *S. typhi* Ty 21a. One live attenuated cholera vaccine product has been well tolerated and demonstrated to be immunogenic in adults. Clinical trials of this vaccine in children are in progress.

Influenza

Influenza virus infection continues to cause significant morbidity and mortality despite the availability of effective vaccines and antiviral therapy for prevention and treatment of influenza. Influenza viruses cause annua epidemics of acute respiratory disease that affect all ages.

A major difficulty in the development and provision of satisfactor immunizing agents for the prevention of influenzal disease is the antigeni variation in these viruses. Periodic minor antigenic changes in influenza A or B virus is the major factor in the continuing occurrence of yearly influenza disease. While outbreaks are generally limited in magnitude, resulting mot bidity and mortality remains discouragingly high. Major antigenic changes o influenza A virus, as occurred in 1957 to 1958 (Asian strain) and again ii 1968 to 1969 (Hong Kong variant), account for pandemic spread of diseas associated with greater overall morbidity and mortality, especially in high risk populations.

Trivalent inactivated vaccines are the only currently available vac cines against influenza viruses in the United States. The killed vaccin must be given yearly because antibody response is short-lived and because of the yearly antigenic variation in the circulating strains of in fluenza. The vaccine efficacy is variable, depending on the immune response to vaccination and how closely the vaccine strain resembles the circulating strains of influenza. To improve the immunogenicity of inactivated influenza vaccine, several new approaches are under investigation. These include production of hemagglutinin in *Escherichia coli* by means of recombinant DNA technology; synthesis of parts of the hemagglutinin molecule in yeast cells; conjugation of the hemagglutinin to diphtheria toxoid; incorporation of the genes for hemagglutinin and neuraminidase into vaccinia virus and use of more potent adjuvants.

Live attenuated influenza virus vaccines are also under development. The attenuated influenza virus vaccines are reassortants containing two RNA genes encoding the two surface glycoproteins, hemagglutinin and neuraminidase, that are derived from the currently circulating virus and include six genes encoding for the internal viral proteins from an attenuated strain. The two types of attenuated virus that appear most promising for clinical use are cold-adapted and avian influenza strains.

Cold-adapted strains are produced by growing influenza A virus at progressively lower temperatures to produce a strain that grows well at 25-33°C but not at 37°C. This property allows the vaccine strain to replicate in the upper respiratory tract but restricts its replication in the lower respiratory tract. One such cold-adapted strain, A/Ann Arbor/6/60 (H2N2), that has genetically stable mutations in the six genes coding for the internal proteins of the virus has been evaluated in human trials. In these trials, coldadapted Ann Arbor Influenza A vaccines given at the site of natural infection, the respiratory mucosa, produce little or no clinical illness, induce serum antibody that persists for at least a year and produce a nasopharyngeal IgA response. However, seroconversion rates have been low, ranging from 56 to 84% of young adults and 73-100% of children. Long-term efficacy trials are in progress. Clinical trials of a cold-adapted influenza B vaccine and trivalent cold-adapted vaccine are currently beginning.

Avian-human recombinant vaccines are made with avian strains which do not cause clinical illness in humans. These vaccines are also undergoing clinical trials.

Pneumococcal Infection

Streptococcus pneumoniae is a leading bacterial pathogen, especially among young children, the elderly and persons with certain high-risk conditions. In children, pneumococci are the most common cause of otitis media, occult bacteremia and bacterial pneumonia requiring hospitalization, and an important cause of bacterial meningitis. In some populations, such as Alaskan Natives, the incidence of bacteremia is markedly higher than that reported in other geographic areas of the United States. Other high-risk groups include children with sickle cell anemia, asplenia, Hodgkin's disease, congenital humoral immunodeficiency, symptomatic HIV infection, nephrotic syndrome, and recipients of organ transplants. Other chronic diseases associated with increased risk of severe pneumococcal disease include chronic cardiovascular and pulmonary diseases, diabetes mellitus and renal failure. The role of these chronic diseases in predisposing to pneumococcal infection has been primarily demonstrated in adults. Mortality is highest in those who have bacteremia or meningitis, the elderly, and in patients with impaired humoral immunity or certain chronic diseases.

Morbidity and mortality of pneumococcal infection appears to be particularly high in developing countries. A review of priorities for vaccine development by an Institute of Medicine committee concluded that the development of an effective pneumococcal vaccine for infants warrants highest priority. The recognition of penicillin-resistant pneumococci further underscores the need for effective pneumococcal vaccines.

The benefits of the currently available pneumococcal capsular polysaccharide vaccines in childhood is limited by their poor immunogenicity in children less than 2 years old. However, pneumococcal polysaccharide-protein conjugate vaccines that are immunogenic in infancy are in development and provide promise for more effective pneumococcal immunoprophylaxis. These vaccines are produced by chemically linking a capsular polysaccharide of pneumococcus to a carrier protein such as diphtheria or tetanus toxoid. As with the *H. influenzae* conjugate vaccines, they are T-cell-dependent antigens which elicit antibody responses in infancy and also booster responses.

One potential problem with such vaccines is the need to immunize against many different serotypes of pneumococci. Because of local reactions to the protein component, conjugate vaccines which contain more than 6-8 serotypes may be difficult to produce. As a result, different formulations of conjugate pneumococcal vaccine may be developed that would contain different serotypes targeted to a specific group of patients. Vaccine containing types 4, 6, 9, 14, 18, 19F and 23F would be necessary for prevention of otitis media in the United States while types 1, 2 and 5 would need to be added to prevent pneumonia in developing countries.

DISEASES FOR WHICH INVESTIGATIONAL VACCINES HAVE BEEN DEVELOPED

Routine immunizations for children have virtually eliminated many infectious diseases from the United States. These successes have encouraged research into vaccines to prevent other serious viral and bacterial diseases affecting children. A 1985 report by the Institute of Medicine and the National Academy of Sciences reviewed the benefits to the United States that would be associated with the development and use of new and improved vaccines. The report listed twelve diseases for which vaccines were possible and where for each the direct costs associated with treatment exceeded $100 million a year. Diseases affecting children targeted for which vaccines are not available included cytomegalovirus (CMV), group B streptococcus, herpes simplex, parainfluenza viruses, respiratory syncytial virus, rotavirus and hepatitis A. Investigational vaccines currently undergoing human trials include those against each of these diseases, except parainfluenza viruses.

The approaches to development of vaccines against the childhood pathogens are varied and reflect the advances in scientific technology. For example, herpes simplex and respiratory syncytial virus vaccines consisting of glycoprotein subunits produced by conventional or recombinant techniques are being tested in target populations. A protein-conjugated polysaccharide vaccine for group B streptococcus has been produced and is undergoing trials for immunogenicity. Live attenuated viral vaccines under development include CMV, hepatitis A, herpes simplex and rotavirus vaccines. CMV and hepatitis A vaccines have been produced by conventional means of attenuation and trials are underway, although development of CMV vaccine has been slowed by concerns over possible oncogenicity of vaccine virus and the potential for reactivation causing perinatal symptomatic CMV infection. Trials of herpes simplex virus vaccine produced by genetic rearrangement are currently beginning. The most promising rotavirus vaccines are reassortants containing one RNA gene encoding the VP7 protein which determines neutralization, derived from the four human serotypes of rotavirus and ten genes from an animal strain which produced attenuated illness in humans. Large-scale trials of immunogenicity and efficacy of these vaccines in children are in progress.

These vaccines and others, if proven effective, will be added to the armamentarium of the pediatrician of the future.

13

INFECTIONS IN DAY CARE CENTER

Since their establishment in the 19th century, there has been a significant increase in utilization of day care centers (DCCs). In the US, the number of licensed DCCs has tripled in the past 15 years, and it has been estimated that approximately 90% of families use out-of-home child care on a full or part-time basis, a phenomenon related to a diversity of reasons, including an increasing number of women who have joined the labor force to seek financial support and to develop professional careers. Other factors include the increasing number of single-parent families and the mobility of young families, making local extended family members less available for child care than in previous decades. Out-of-home child care is usually provided in DCCs and day care homes. In the US the mean enrollment in center-based programs is 68 children, while the mean enrollment of nonresident children in home-based programs is 6 children.

Growth in the number of day care facilities has generated greater awareness of the social and health aspects associated with out-of-home child care. Public health authorities, health care professionals, day care providers and parents are constantly seeking methods to prevent accidents, injuries and child abuse related to day care, while providing a safe environ-ment and promoting optimal nutrition, development and social adjustment. National standards and regulations to ensure quality day care vary in different countries. In the US, the American Academy of Pediatrics and the American Public Health Association have developed national standards for infectious disease control in the child care setting.

Infectious diseases are among the most frequent, important and compli-cated problems encountered in DCCs. The purpose of this manuscript is to review the epidemiology of infections in DCCs, including antibiotic use and resistance. Infections of specific systems will be reviewed, followed by a discussion of preventive and control measures that are necessary to mini-mize disease in DCCs.

EPIDEMIOLOGY

Incidence of Infections in Day Care Centers

Although current knowledge is incomplete, several prospective studies have shown that the incidence of infections among children attending DCCs is higher than among age-matched children who are cared for in their homes or in home care facilities. In a prospective study in Pittsburgh, Pennsylva-nia, Wald et al. found that infants in DCCs had more respiratory tract infections during the first year of life. Likewise, Swedish children in DCCs had a higher rate of respiratory infections than children in home care. In US children less than 5 years of age, 10% of respiratory tract illness were attributable to day care attendance. In Finland and Denmark, DCC children had a greater number of episodes of otitis media. These children also had more severe illnesses and more hospitalizations for myringotomy and venti-lation tube placement.

The risk of primary diseases due to *Haemophilus influenzae* type B was significantly higher in young children in DCCs than among children in home care, independent of other risk factors. Infectious gastroenteritis, hepati-tis A and cytomegalovirus infections are also highly prevalent in DCCs. Acute infectious diarrhea is twice as common in diapered children in day care than in children cared for in their homes.

Spread of Infections in DCCs

Infectious agents are readily transmitted among children in DCCs. Outbreaks of diarrhea or respiratory infections are common. Transmission of organisms is related to the conditions in the DCCs and the behavior of the children. Overcrowding is often seen, with too many children of different ages in small rooms. Food preparation and eating areas are sometimes not separated. In addition, infants who are not toilet-trained may cause fecal contamination of toys, furniture and other environmental objects. Coughing and sneezing at close range is common, and may result in the spread of respiratory tract pathogens. Children often tend to put contami-nated hands or objects into their mouths. Because most infections are communicable for a few days before and a few days after signs of disease are apparent, infection may spread before it is recognized in the index case. This is particularly common in infections due to hepatitis A virus in which outbreaks in DCC are detected when adult contacts become symptomatic following infection.

Timing of Infections in DCC Children

Prospective studies have shown what seemed logical: the rate of infec-tious disease is highest in children who were newly enrolled in the DCCs. A follow-up of 442 children younger than 2 years who attended DCCs in Houston, Tex., showed that diarrheal incidence rates were higher in the first 4 weeks in children in DCCs than in subsequent weeks (4.4 vs. 2.7 cases/ child-year, $p<0.01$) [A 3-year prospective study showed that the increased incidence of respiratory tract infections in DCC attendants tend to cease with time. Primary infections with *H. influenzae* type B were 9 times higher in children less than 2 years of age during their first month of attendance than in age-matched children enrolled for longer periods of time.

Transmission of Infections from DCCs to Adults

Emphasis has been placed on the importance of transmitting pathogens from children in DCCs to their providers and family members. Infections that are commonly acquired by adults from children in DCCs include gastrointestinal tract pathogens *(Shigella sp., Giardia* and rotavirus), respira-tory tract pathogens (influenza, parainfluenza and rhinovirus), hepatitis A, herpesviruses (cytomegalovirus and herpes simplex), parvovirus B 19 and skin infections (impetigo, lice and scabies).

ANTIBIOTIC USE AND RESISTANT ORGANISMS

Antibiotic Use in DCCs

The increase in infectious diseases, especially otitis media, in children in DCCs results in an increase in the use of antimicrobial agents. Several controlled studies showed increased antibiotic use in children attending DCCs as compared to age-matched controls with no care outside their homes. In Finland, the calculated annual rates of antibiotic use per child-year in day care were two times higher than children in other settings. A detailed study in Houston showed that among children attending DCCs, 36% received antibiotics during an 8-week period, versus 7% and 8% among children in day care homes or those cared for at home, respectively ($p<0.001$) (table 1). The calculated annual rates of antibiotic use were similar for children in day care homes or cared for at home, but was 3.6 times higher for DCC attendees, who also required significantly longer courses of antibiotics.

TABLE 13.1. ANTIBIOTIC USE IN CHILDREN UNDER 3 YEARS OF AGE DURING AN 8-WEEK STUDY

Types of Child care	*Children receiving antibiotics*	*Average days of antibiotic use per chile*	*Annual rates of antibiotic use per child-year*
Day care centers	35.7	19.9	9.1
Day care homes	7.1	4.0	2.7
Cared at home	8.2	4.6	2.5

Resistant Microorganisms

Various reports of single cases and outbreaks suggested an association between DCC attendance and resistant strains *of Streptococcus pneumoniae. H. influenzae, Escherichia coli* and *Shigella sonnei*. Controlled studies evaluating antibiotic resistance in DCCs are limited.

Among respiratory pathogens, strains of *S. pneumoniae* that were resis-tant to penicillin, trimethoprim/sulfamethoxazole ('TMP/SMX) and tetra-cycline, caused invasive disease in children in Colorado and Houston. In both cases, an investigation showed that 12-24% of the classmates of the index cases were colonized with similar resistant strain(s). These studies documented that colonization with resistant S. *pneumoniae* was common among children in DCCs, but there was no comparison group. In South Africa, resistant strains were more common in children in UCCs than in other children. Likewise, several reports described invasive disease due to resistant *H. influenzae* among children attending DCCs, but without a control group. A prospective study in Canada showed that 8% of children attending DCCs and 2% of children not in DCCs were colonized with beta-lactamase producing *H. influenzae,* but the difference did not reach a statistical significance.

Among enteric pathogens, 2 outbreaks of multiple-resistant *S. sonnei* in DCCs have been reported, but it is not known if antimicrobial resistant *Shigella* strains are more prevalent in DCCs in Houston. Fecal colonization tans EE *coli* has been studied in comparative studies in DCCs in Houston. Multiple-resistant *E. coli* were more common among diapered children under 2 years of age wino attended DCCs than among other children of (31 vs. 5-8%, $p < 0.001$). Only the number of diapered children atte center and younger age were correlated with antibiotic resistance. R strains were usually carried for months and in 25% of the cases, they transferred to family members, usually mothers and siblings. This e sizes the possibility that diapered children in DCCs may serve as a sou resistant strains that may spread to the community.

INVASIVE INFECTIONS IN DCCS

Because of their severity, potential morbidity and mortality, sen infections in DCCs are the subject of considerable and increasing concern. *H. influenzae* type b is the major cause of invasive bacterial infections in children younger than 5 years of age, including bacteremia, meningitis, septic arthritis, pneumonia, epiglottitis, cellulitis and pericarditis. Because age is a major determinant affecting *H. influenzae* type B disease, age-matched controls are mandatory in epidemiologic studies related to infection rates.

Invasive bacterial disease in a day care setting has been categorized as being either primary or secondary (subsequent). Primary disease occurs in the absence of known contact with another case in the center in the previous 60 days. Several controlled studies in US and Europe have examined the relative risk of primary *H. influenzae* disease in children who attended DCCs as compared to children who did not. Although studies have found a significantly increased risk related to day care attendance, the magnitude of the increase differed, ranging between 1.8 and 19.6. The differences are related, in part, to methodology, the age groups being studied, geographic location and the size of the DCC. Generally, the younger the children, the higher the relative *risk*.

A controversial issue concerns secondary or subsequent disease, which is defined as the *risk* of *H. influenzae* disease among children attending DCCs who are exposed for 4 h or longer to an attendee with disease within 60 days. In such studies, the risk of acquiring the disease from DCC contacts ranged between 0 and 3.2%. Temporal and geographic variations in the risk probably exist. With the current use of *H. influenzae B* vaccine in children beginning at age 2 months, the occurrence of this organism in the day care setting should dramatically decline.

Neisseria Meningitides

Reports have described secondary cases of *N. meningitides* disease in DCC settings. A review of cases reported in 1971-1976 showed that the risk of *N. meningitides* disease among DCC contacts of an index case was about 4/1,000, which was 75 times higher than in the general population. In a report on group day care and risk of serious infectious illnesses, data suggested an association between day care attendance and invasive *N. men-ingitides* disease (OR 2.0, CI 0.39-10.27), but the correlations were not significant.

RESPIRATORY TRACT INFECTIONS

Acute respiratory tract infections are among thee most common problems encountered in young children. Respiratory tract infections are usually classified according to the anatomic site of involvement.

Upper Respiratory Tract Infections

Prospective studies in Europe and the US have shown that upper respiratory infections, including acute or chronic otitis media, are more common among children attending DCCs than among age-matched controls. The incidence is highest in young children, especially during the first year of life. Children attending home care had lower rates of infection. Compared to their peers who do not attend DCCs, children less than 2 years of age who attend group day care experience about 50% more episodes of otitis media. Because of these illnesses, children in day care receive signifi-cantly more antibiotics for longer periods of time than children not in day care.

Lower Respiratory Tract Infections

In contrast to upper respiratory tract infections, incomplete data are available regarding lower respiratory tract infections in day center settings. A prospective study in Sweden found no difference in the frequency of lower respiratory tract infections among children in DCCs, family day care homes and home care. By contrast, in a study based on hospital admissions for lower respiratory tract infections of children under 2 years of age, DCC attendance increased the risk of infection by about twofold. In a study evaluating serious infectious illnesses in children, group day care attendance did not pose a risk of serious infections, other than those due to *H. influenzae* and *N. meningitidis,* to young children.

ENTERIC INFECTIONS IN DCCS

Diarrheal illness in DCCs often presents as outbreaks, most likely because of the ease of person-to-person transmission of many enteropathogens, especially among diapered children. This results from fecal contamination of the environment by these children and need for close handling by the staff. The incidence of diarrheal disease among infants and toddlers in DCCs has been shown to be significantly higher than among age-matched children it day care homes or those cared for at home. Three well conducted studies have compared the risk of diarrheal illness among children in DCCs, day care homes and home care. These studies have shown the relative risk of diarrhea in children in DCCs to be 2.2-3.5 times higher than in children in home care. In only one study was the relative risk higher (2.0) in children in day care homes than children cared for at home. The incidence of diarrhea in diapered children in day care is approximately 3 episodes/child/year.

Because of their low infectious dose, *Shigella sp.* have been recognized as the major bacterial enteropathogen in DCCs, documented both as sporadic cases in longitudinal studies and, more typically, as minor or major out-breaks. Other bacteria also prevalent, well documented and important include *Campylobacter sp., Clostridium difficile* and various groups of diar-rhea-associated *E. coli*. The latter include outbreaks of enteropathogenic *E. coli* (EPEC) that are characterized by the chronic or relapsing nature of the diarrhea. Severe outbreaks caused by serotypes 011 1:K58 or 0114:NM, with dehydration and prolonged hospitalization of some of the infected infants, have been reported. In addition, diarrheal outbreaks due to cytotoxin-producing *E. coli* serotype 0157:H7 have been recognized. Outbreaks with this group of *E. coli* were associated with development of the life-threatening hemolytic-uremic syndrome. Salmonella gastroenteritis is rare in DCCs.

Outbreaks of rotavirus diarrhea in DCCs are common. They are usually encountered during the colder months of the year and involve diapered children. In most outbreaks a single serotype of rotavirus has been involved, although sometimes multiple serotypes, or even multiple enteropathogens, were identified. Recurrent infections by the same serotype have been documented by a longitudinal study in DCCs in Houston.

With the increasing use of electron microscopy and the development of new detection methods, other viral causes of gastroenteritis have been documented, including enteric adenoviruses, calcivirus and astrovirus.

The two parasitic organisms that commonly infect children in DCCs are *Giardia lamblia* and *Cryptosporidium*. *G. lamblia is* the most common parasitic infection encountered in DCCs. Using the simpler and more sensitive antigen detection assays, diagnosing *G. lamblia* infections has become less complicated, making epidemiologic studies easier to conduct. The attack rates during outbreaks of *G. lamblia* can be as high as 50% and infection usually occurs in infants and young children. During the outbreaks, excretion of the organism by asymptomatic children frequently occurs. In one study, stools that were collected weekly during a 15-month longitu-dinal study and tested by ELISA showed that *G. lamblia* infection occurred in 33% of the children in a DCC. Many children had more than one episode of infection and 78% had asymptomatic infection. The mean duration of excretion of the pathogen was 2 weeks. It has even been suggested that asymptomatic children infected with *G. lamblia* had a better weight and height gain and fewer gastrointestinal symptoms than noninfected children.

Initially recognized as a pathogen in immunocompromized patients, *Cryptosporidium* has been documented as a cause of diarrhea also in immunocompetent patients. Outbreaks of cryptosporidiosis in DCCs have been reported, with documentation of asymptomatic excretion of the pathogen by some children.

Hepatitis

Among the various causes of viral hepatitis, hepatitis A virus, which is transmitted by the fecal-oral route, is the most common cause of hepatitis outbreaks in DCCs. Outbreaks have been reported from many locations, sometimes being the source of large community outbreaks. About 14% of cases of hepatitis A reported to the Centers for Disease Control were either a child, an employee in a DCC or their contacts. Typically, infection among DCC children is mild and nonspecific or asymptomatic, and out-breaks may be recognized when adult contacts of the children, often their parents, become ill. Risk factors for hepatitis A outbreaks were studied in a 2-year study. Outbreaks were more common in centers that were large, open for longer hours and enrolled children younger than 2 years of age.

OTHER INFECTIONS

Hepatitis B is transmitted by exposure to infective blood or other body fluids. Since person-to-person transmission is well documented in situations of close personal contact for a prolonged period, the potential for transmis-sion of hepatitis B in DCCs exists. Only few reports have documented such transmission. Because of limited data, the risk of hepatitis B transmission in day care settings cannot be precisely quantified, but it seems to be low. Data on the transmission of hepatitis C, D or E in the day care setting is usually not available.

Cytomegalovirus (CMV)

Several studies support the concept that DCCs serve as an important reservoir of CMV, with frequent transmission both within day care and to the community. Excretion of the virus is higher among DCC attendees than among matched controls. The percentage of children excreting CMV has been shown to increase from about 10% during. the first year of life to about 60-80% during the second and third years of life. This suggests spread among children in the DCC, as has been documented by restriction endonuclease analysis of DNA of CMV isolates.

Several studies have documented that CMV excreted by young children was an important source of infection for women who were in close contact with them, namely their mothers and care providers. The incidence of CMV infection of day care workers as measured by seroconversion in 3 studies ranged from 8 to 20% per year. Controls for these studies had an annual seroconversion rate of 2%. The importance of this transmission is in its ability to cause congenital CMV infection with multi-system involvement and severe sequelae in susceptible hosts. Infection of immunocompromized hosts is also a serious complication, associated with considerable mortality and morbidity.

Human Immunodeficiency Virus

With the increasing epidemic of HIV infection and AIDS in children, concern regarding the risk for and from these children in the day care setting has been raised. Regarding the risks of day care for the HIV-infected child, there is a danger of being infected with bacterial or opportunistic organisms because of defects in humoral and cell-mediated immunity. Children with HIV infection in the day care setting need to be monitored for exposure to infectious diseases and have their health and immune status frequently evaluated. There is no evidence of HIV transmission within the day care setting by an infected child.

Parvovirus B19

In healthy children parvovirus B 19 caused outbreaks of erythema infectiosum or fifth disease. The significance of parvovirus B 19 as a pathogen is in its ability to infect erythroid precursor cells. In immunocompromized per sons it can produce severe, chronic anemia and in susceptible pregnant care setting is women infection may result in hydrops fetalis or abortion. Congenital malformations have not been attributed to human parvovirus B 19 infection. Serologic evidence of previous infection has been reported in 30-60% of adults and 2-15% of preschoolers, indicating that infection is most often Milt acquired in early childhood. Transmission is through the respiratory route. d to Attack rates of 36% and 38% have been reported among susceptible health care workers exposed to patients with transient crisis. Studies of g outbreaks in schools and DCCs have shown that parvovirus is readily life transmitted from child to child and from child to adult. In outbreaks of is erythema infectiosum in schools, 10-60% of children developed a rash and a seroconversion occurred in 31 % of day care personnel. In a study of risk factors in pregnant women in a community with an outbreak of erythema infectiosum, the highest rates of infection were found in school teachers (16%) and day care workers (9%). Fortunately, fetal infection was relatively uncommon even during community outbreaks.

Appropriate prevention and control measures may have a significant y effect on the rate of infections within the DCC settings. Education of care providers regarding appropriate personal and environmental hygiene is of prime importance. Handwashing is probably the most important infection control practice and has been shown to reduce certain infections, especially gastroenteritis, in DCCs.

The physical space, organization and the quality of the environmental facilities of day care premises also are relevant. Overcrowding should be prevented and care provider to child ratio should be appropriate, adhering to the national standards. Unfortunately, in the US it appears as if centers, are serving more children with fewer workers. Older children should be separated from infants. Diaper changing areas should never be located in food preparation areas, and should never be used for temporary placement of food. Diaper changing surfaces should be nonporous. Toilet-training equipment should be maintained in a sanitary condition. Each item of sleep equipment should be used only by a single child while enrolled, and should be cleaned and sanitized prior to assignment to another child.

Although exclusion of staff and children with suspected or proven infectious diseases is mandatory, proper implementation of this principle is problematic. Mild infections are common among young children and it would be undesirable to exclude them from their usual base of care for extended periods. Moreover, excretion of the infecting agent is often seen for some period before and after the symptomatic illness and sometimes even without any clinical diseases. These problems were addressed by experts of the National Child Care Standards Projects. They recommend exclusion for illnesses known to be transmitted among children and in cases of serious illnesses, even when the risk of transmission is hypothetical and there are insufficient data to quantitate the risk. Children with minor illnesses need to be excluded if: (1) the illness prevents the child from participating comfort-ably in program activities; (2) the illness causes a need for more care than the staff can provide without compromising the health and safety of the other children; or (3) the child has certain infectious diseases (pertussis, mumps, hepatitis A, diarrhea, etc.).

During outbreaks, cohorting (inclusion of infected children in a group with separate staff and facilities) and closure of the center for

TABLE 13.2. MAIN MEASURES TO REDUCE INFECTIONS IN DAY CARE CENTERS

1. Education of care providers, mainly on
   - a Importance of personal and environmental hygiene, especially hand washing
   - b Supervision of the use of the toilet
2. Adherence to day care center standards with regard to
   - a Preventing overcrowding
   - b Separation of older children from infants
   - c Separation of food-preparation and eating areas
   - d Having appropriate care providers/children ratio
3. Exclusion of staff and children with suspected or proven infectious diseases 4 During an outbreak
   - a Cohorting of infected children (if possible)
   - b Closure of the center for new enrollments
   - c Use of appropriate pre- and/or postexposure chemoprophylaxis or immunization
5. Verification of immunization coverage of attendees and staff
6. Development of infectious surveillance systems

new enrollment are other considerations. According to the specific disease or microorganisms involved, active or passive immunization or chemoprophylaxis may be neces-sary. In particular, antibiotic therapy may shorten the shedding period of certain microorganisms, sometimes without affecting the clinical course of the disease. Proper use of such antibiotics may be important in the day care setting. Routine immunization is of particular importance in children in DCCs; all children enrolled in a DCC should provide written documentation of up-to-date immunizations.

CONCLUSIONS AND FUTURE TRENDS

DCC settings provide an environment in which infectious disease can spread among children and sometimes even be transmitted to the community. Frequent antibiotic use and development of multiple-resistant strains may become an increasing problem in the future, hampering appropriate therapy. Appropriate hygienic measures, often as simple as handwashing, may stop the chain of transmission. With our better understanding of pathogenesis, epidemiology and cellular mechanisms of some infections and improved methods of diagnosis, prevention of certain infections may become possible, by providing efficacious vaccines or other control measures.

14

Febrile Child

Fever in pediatric patients younger than 24 months of age is common and plays an important role in the daily activity of practicing pediatricians both in the community and in the hospital setting. However, the febrile young patient still poses one of the most vivid and emotional dilemmas to the pediatrician and controversies are still present. When the young pediatric patient looks seriously ill, there is little doubt as to the management. However, it is now clear that a small proportion of febrile patients younger than 24 months of age are bacteremic despite their relatively well appearance, and that a small proportion of this subgroup will go on to develop serious complications such as meningitis or septicemia.

The main questions that arise when a febrile young patient presents to the pediatrician are: (1) Can we detect those who are bacteremic or at least those who are at increased risk of being bacteremic? (2) Should those children who are at increased risk of being bacteremic be treated with antimicrobial agents? (3) Which is the most cost-effective way to manage such febrile patients younger than 24 months of age?

In this chapter we will review the natural history, epidemiology and the controversies regarding the management of febrile, previously healthy young children and infants, and propose a scheme for the management of these patients.

THE FEBRILE INFANT YOUNGER THAN 3 MONTHS OF AGE

Only 1-2% of all children present to the physician with a significant fever (>38.3 °C) during the first 3 months of life. However, the proportion of infants who have serious infectious diseases among this group is substantially greater than among older pediatric patients. In one study conducted at the Yale-New Haven Hospital, 8/22 (36%) of infants younger than 3 months with a temperature of > 103 °F had a serious disease. It was found that clinical impression combined with history, physical examination and laboratory tests was quite sensitive in predicting the presence of serious disease. However, all the tested combinations did not point out accurately enough the patient with bacteremia or meningitis in this age group. Accuracy is decreased if the physician is less experienced and if the child is less than I month old. During the immediate postnatal period (particularly the first week of life) the risk is maximal, especially for hospitalized premature babies.

Infectious causes of fever in young infants vary with age, but most cases are caused by viral pathogens. The main organisms that cause bacteremia and bacterial meningitis during the first 2 months of life are group B streptococcus, enteric gram-negative rods, *Listeria monocvtogenes, Streptococcus pneumoniae* and staphylococci. After the first 3 months of life, the most important organisms are *Haemophilus influenzae* type b, *S. pneumoniae, Neisseria meningitidis* and *Salmonella spp.*

Because of the inability to identify the febrile infants with bacteremia or meningitis and because of the wide spectrum of involved bacteria, most pediatricians in academic institutions prefer to hospitalize such infants, to obtain blood, cerebrospinal fluid (CSF) and urine cultures and to start a broad-spectrum parenteral antibiotic regimen for at least 48 h, until culture results are known. However, hospitalization may cause unnecessary complications: in one study conducted at the Department of Pediatrics at the Johns Hopkins University, 19.5% of all admissions of febrile infants <60 days of age resulted in at least one complication.

It soon became apparent that although protocols in many academic institutions demanded measures as mentioned above, many physicians including private practitioners and emergency room physicians did not adhere completely to the protocol.

In 1985, Dagan et al. published a new approach for febrile infants younger than 3 months of age: instead of trying to identify the febrile infants who have serious bacterial infection /SBI, they tried to identify the febrile infants younger than 3 months of age who were *unlikely* to have SBI. Previously healthy infants were considered at low risk for having a SBI if they had no findings consistent with soft tissue, skeletal or ear infection, a peripheral white blood cell count (WBC) between 5,000 and 15.000/mm^3. an absolute band count of < 1,500/mm^3 and a urinalysis which had < 10 WBC/ high power microscopic field in centrifuged sediment. In subsequent studies, infants with diarrhea were considered at low risk if there were fewer than 25 leukocytes on fecal smear. Such criteria had a negative predictive value of 99.53% with a 95% confidence interval of 98.14-99.92% for SBI and 99.77% with a 95% confidence interval of 99.50 to 99.99% for bacteremia.

Two prospective studies have suggested that careful outpatient management of infants considered to be at low risk for having SBI may be safe. Dagan et al., in Israel, observed most low risk infants as outpatients without antibiotic treatment and McCarthy et al., in Rochester, N.Y., managed such infants with intramuscular injections of ceftriaxone once daily until culture results were available. In one study no lumbar puncture was performed and no CSF or urine were cultured if the child was considered at low risk for having bacterial infections according to the previously described criteria. In addition, the results of a retrospective study by Wasserman and White on 443 consecutive febrile infants younger than 3 months suggested that many febrile infants who are hospitalized for observation can be managed as outpatients.

THE FEBRILE CHILD AGED 3-24 MONTHS

In this age group, the overall rates of bacteremia that were reported in many studies ranged from 3 to 8%. The rate of bacteremia without apparent focus among these children was in the range of 2-5% whether the child was seen in the hospital or in the clinic. The controversy regarding diagnostic tests that predict bacteremia and the need for antibiotic treatment in febrile children aged 3-24 months is even greater than those in younger infants, despite the fact that fever is extremely frequent in this age group.

The most common organisms to cause bacteremia and meningitis are bacteria that contain a polysaccharide capsule and frequently colonize the upper respiratory tract, namely *H. influenzae* type b, *S. pneumoniae* and *N. meningitidis*. It is probable that in the near future,

H. influenzae type b will become an increasingly rare cause of bacteremia, since highly efficacious vaccines against these organisms have lately been introduced.

It was not until about 30 years ago that pediatricians started to observe that bacterernia may be present in some young febrile children who do not appear toxic and who do not have an apparent focus. This entity was named 'occult' bacteremia, 'silent' bacteremia or 'walking' bacteremia. In a large series, Shapiro et al. observed that 21 % of 1,483 bacteremic cases that were seen at two major hospitals had occult bacteremia: 44% of all cases with pneumococcal bacteremia were 'occult' versus 20% of all cases with meningococcal bacteremia and 8% of all cases with *H. influenzae* type b bacteremia.

Most of the children with occult bacteremia recover without developing a serious complication, and this is especially true for pneumococcal bacteremia. However, some of the children with occult bacteremia will develop focal complications, including meningitis. The risk to develop the most dreadful complication, namely meningitis, is not similar when the 3 main pathogens are compared: compared to the risk of meningitis in cases with pneumococcal bacteremia, patients w!.Th *Hae mophilus* bacteremia had 15 times and those with meningococcal bacteremia had 81 times greater chance to develop meningitis. Chifortunateiy, to date, no test can predict those infants who will have serious complications.

Numerous investigators have tried to identify the febrile children aged 3-24 months with a higher risk of having bacteremia. The main diagro5tic tests that have been investigated are clinical scoring systems, WRC' and differential, microscopic examination of the buffy coat of the blood, erythrocyte sedimentation rate, c-reactive protein, morphologic changes in neutrophils on a stained smear of the blood, and quantitative cultures of the blood. All the positive tests were associated with increased risk of having bacteremia, but none, single or in combination, were able to predict those patients with bacteremia. In other terms, the sensitivity and positive and negative predictive value were insufficient when all tests were studied.

In view of the frequency of occult bacteremia among febrile patients younger than 24 months of age without an apparent focus, and the difficulty to identify those who will develop serious complications, it is easy to understand why management is controversial.

One question is whether to obtain blood cultures from all such patients. Blood culture is the only definitive means to diagnose a bacteremic child. However, cultures do not yield results until at least 24 h after the initial visit, during which time children may develop focal infection, or may be lost to follow-up. Furthermore, the risks of the procedure are pain, unnecessary hospitalization and intravenous antibiotic treatment in children with a positive blood culture who are still febrile but whose bacteremia has spontaneously resolved.

Another question is whether to administer empiric treatment to all febrile children with oral or parenteral antibiotics pending culture results. Antibiotic treatment to all febrile children should reduce complications from the infection. However, empiric antibiotic therapy to a group in which most patients will not have bacteremia will expose many children to avoidable risks of therapy. Clinical judgement or WBC may help to select patients at high risk for the presence of bacteremia, but treatment of only high risk patients will still leave a proportion of untreated patients at risk for the development of serious complications.

The discussion about the need to treat empirically tends to get more complicated since the available results do not show a clear advantage of early treatment when the most serious complication of occult bacteremia, namely meningitis, is considered. Data are available from 7 comparative studies in which 460 children with occult bacteremia with *S. pneumoniae, H. influenzae* type b or *N. meningitides* were enrolled; 265 children were treated and 195 children were not. Although the study design and antibiotic agents were not identical in all studies, the results suggest a clear advantage of early treatment in eliminating bacteremia (3.4% were bacteremic at the next follow-up visit in the treatment groups versus 16.9% in the untreated children) and prevention of focal infection (1.9% of treated patients had an apparent focal infection at the next follow-up visit versus 14.9% of the untreated patients). In contrast, no apparent advantage was observed in the prevention of meningitis: 5.7% of the 265 treated patients developed meningitis versus 6.7% of the 195 untreated patients.

In view of the intriguing results, Lieu et al. have tried to use a decision analysis to evaluate the probable health benefits, complications and costs of six management strategies: (1) no intervention; (2) antibiotic treatment alone: (3) blood culture alone; (4) blood culture plus antibiotic treat ment; (5) screening by WBC (blood culture + treatment for those with > 10,000 WBC/mm'), and (6) clinical judgement (blood culture + treatment for high risk). Treatment consisted of oral amoxicillin therapy for 2 days. The strategy that combined blood culture with empiric oral antibiotic treatment of all patients was predicted to prevent the highest number of major infections and to have the lowest cost per major infection prevented. The strategy that combined WBC and blood culture for all patients, followed by empiric antibiotic treatment for those with leukocyte count > 10,000/mm', had almost equal cost and clinical effectiveness and still prevented many antibiotic complications. Blood culture without empiric treatment constituted the third most clinically effective intervention but was the least costeffective method in this model. Giving a 2-day oral course of amoxicillin without testing had the lowest average cost per febrile patient but was the least clinically effective intervention. However, the low degree of effectiveness of empiric treatment alone was based on the assumption that oral amoxicillin therapy was only 20% effective in preventing major infection after bacteremia. At higher estimates of effectiveness, treatment alone became a more viable strategy. However, it must be remembered that the assumptions in such an analysis are based on data which may differ in various geographical and ethnic settings and thus may not be uniformly valid worldwide. Furthermore, with the introduction of successful conjugate *H. influenzae b* vaccines, occult bacteremia will be less frequent and may be less dangerous since the most common cause of bacterial meningitis in the Western World *is H. infuenzaeb*. The available data support, in our opinion, the management scheme described in figure.

The choice of antibiotics *is* also inconclusive. Acceptable oral drugs are amoxicillin, amoxicillin/clavulanic acid, trimethoprim/ sulfamethoxazole, and cefixime, among others. Some authorities prefer intramuscular administration of ceftriaxone. If antibiotic treatment is started it should probably continue for at least 2 days.

More definitive studies regarding the management in the era of successful *Haemophilus* immunization need to be performed.

15

NEURAL INFECTIONS

BACTERIAL MENINGITIS

Epidemiology

The incidence of systemic infections including meningitis due to *Haemophilus influenzae* type b (Hib) is declining in countries where Hib protein conjugate vaccines are routinely administered to infants. For example, in Helsinki, Finland bacterial meningitis due to Hib has virtually disappeared. Although the most recent published survey by the US Centers for Disease Control (CDC) for 1986 indicated that Hib, *Streptococcus pneumoniae* and *Neisseria meningitidis,* in that order, were the leading causes of bacterial meningitis in infants and children, the relative incidence of these pathogens has clearly changed since the introduction of infant Hib immunization. Serogroups B (48.4%) and C (46.4%) are the most commonly isolated meningococcal strains.

Pathophysiology

The molecular pathogenesis of the inflammatory responses during bacterial meningitis has been the subject of many experimental and clinical studies over the last several years. The sequence of events following the initial invasion of the central nervous system (CNS), by microorganisms is closely tied to the production of cytokines, especially tumor necrosis factor (TNF), interleukin-1 (IL-1) beta and interleukin-6 (IL-6). Lipopolysaccharides released from gram-negative organisms and cell wall components of gram-positive organisms appear to stimulate the production of cytokines by macrophages, monocytes, astrocytes, and microglial cells. These cytokines promote the appearance of neutrophils across endothelial surfaces into tissues. This inflammatory response is thought to result in endothelial cell injury, cerebral edema, increased intracranial pressure and alteration of cerebral blood flow (CBF). In studies conducted in Dallas, concentrations of IL- I in cerebrospinal fluid (CSF) at admission correlated with parameters of CSF inflammation as well as the occurrence of subsequent neurologic sequelae. TNF levels in CSF correlated with CSF protein concentrations, but not outcome. The IL-i concentrations in ventricular fluid from infants with gram-negative enteric meningitis were greater for infants who received systemic antibiotics plus intraventricular gentamicin compared to infants receiving systemic therapy only. Endotoxin and protein levels as well as white blood cell counts were significantly elevated in the intraventricular therapy group. The increased IL-1 concentrations and inflammation in the intraventricular group may help explain the poor outcome in this group compared to systemically treated babies.

CBF has been measured within the first *24-48* h of admission to the hospital in 18 children with bacterial meningitis by a stable xenon computed tomographic technique. Total CBF (62 ± 20 *ml/min/100* g) was considered to be within the normal range for age in 13 (72%) although in many of the patients local areas of reduced flow were present. Five children had reduced total CBF and generally reduced regional flows as well. Many of these children developed severe signs and symptoms of increased intracranial pressure (seizures, 15; respiratory impairment, 10; herniation, 4) and yet total CBF was within normal range in the majority. Thus, the clinical significance of these findings remains to be determined.

Complications

New knowledge regarding the clinical manifestations of bacterial meningitis are basically related to seizures and subdural effusions. Pomeroy et al. reported that the most important predictor of late seizures (epilepsy) was the presence of other neurologic deficits, presumably due to cerebral injuries sustained during bacterial meningitis. In a subset of this group of children between the ages of 1 and *18* months, transillumination was used as a noninvasive screening device for detecting subdural effusions. Effusion developed in 44 (39%) of 113 patients. Although the presence of subdural effusions did correlate with seizures during treatment and neurologic abnormalities still present at the end of treatment, there was no association between subdural effusion and impaired development/intellect or residual neurologic abnormalities at long-term follow-up. There was no evidence that the effusions caused persistent fever in these children. This study supports a noninvasive approach to managing patients with subdural effusion unless a subdural empyema is suspected.

Antibiotic Therapy: Hib and S. pneumoniae

The empiric approach to the antibiotic therapy of bacterial meningitis for children between 3 months and 10 years of age remains either cefotaxime or ceftriaxone alone or the combination of ampicillin and chloramphenicol. Ampicillin plus cefotaxime or ceftriaxone is recommended for children 4-12 weeks of age. In a 5-state survey for the year 1986, the CDC reported that 32% of Hib isolates recovered from patients with meningitis or nonmeningitis systemic infection were resistant to ampicillin.

There is a worldwide problem of decreased susceptibility to penicillin in strains of *S. pneumoniae* (MIC>0.1 μg/ml). In our own institution, a major increase in pneumococcal isolates relatively or completely resistant to penicillin began in 198 . Twenty-two of 172 isolates (12.8%), including 2 from CSF and 4 from blood cultures, were relatively resistant to penicillin. In our case-control study of all pneumococcal isolates recovered from middle ear cultures in 1989, the only factor associated with penicillin resistance was prior administration of amoxicillin-clavulanate.

Penicillin is not adequate therapy for treating bacterial meningitis due to *S. pneumoniae* relatively or fully resistant to penicillin. Optimal therapy is still not established but vancomycin, ceftriaxone, cefotaxime, and chloramphenicol have all been administered 'in such patients. Our experience with cefotaxime has also been favorable in the small number of children we have treated with this type of meningitis. Sloas et al. (ICAAC 1991) have reported 3 treatment failures with cefotaxime or ceftriaxone in children with meningitis due to strains relatively resistant (2) or fully resistant (1) to penicillin. These patients continued to be febrile, remained irritable and 2 had repeated positive CSF cultures several days after therapy had been initiated. Chloramphenicol and vancomycin led to sterile cultures and clinical improvement of these patients.

All pneumococcal isolates recovered from normally sterile sites should be screened for susceptibility to penicillin by Kirby-Bauer disk technique using a 1.0-μg oxacillin disk . Strains with zones of inhibition >20 mm are susceptible to penicillin (MIC<0.I μg/ml). Strains with inhibition zones <20 mm require further testing for documenting penicillin resistance. In general, strains that are fully resistant to penicillin (MIC> 1.0 μg/ml) have no zone of inhibition. Quantitative susceptibility tests indicate that *S. pneumoniae* isolates relatively or fully resistant to penicillin are less susceptible to cefotaxime and other cephalosporins than strains susceptible to penicillin, presumably on the basis of alterations in penicillin binding proteins.

Antibiotic Therapy: Gram-Negative Bacteria

No major changes in the antibiotic treatment of meningitis due to gramnegative enteric meningitis have occurred over the past few years. The only comparative trial did not demonstrate any differences in outcome between a third-generation cephalosporin (moxalactam) and ampicillin plus amikacin . Two studies have reported on the efficacy of cefotaxime alone or in combination with a systemic and/or intraventricular aminoglycoside in a total of 38 infants and children . Children without any underlying defect or conditions predisposing to infection promptly had sterile CSF cultures, whereas children with such conditions required several days to clear their pathogens. Whether the combination of cefotaxime and an aminoglycoside, which may demonstrate synergy in vitro, is superior to cefotaxime alone for treating gram-negative enteric meningitis is unknown. However, I prefer to administer an aminoglycoside (generally amikacin) for the first week with cefotaxime, which is continued throughout the duration of therapy.

Adjunctive Therapy

Suppressing the inflammatory response to infection has been the predominant adjunctive measure evaluated in the management of patients with bacterial meningitis. In two clinical trials dexamethasone (0.15 mg/kg/dose q 6 hx 16 doses) administered 15-20 min prior to or up to several hours after the initial antibiotic dose was associated with a decrease in neurologic sequelae compared to a placebo. In both studies some indices of CSF inflammation and TNF concentrations in CSF were reduced within 12-24 h in the dexamethasone-treated children when compared with placebo adjunctive therapy. In the study conducted in Costa Rica 18 of 48 children (38%) in the placebo group and 7 of 51 (14%) children in the dexamethasone group had one or more neurologic or audiologic sequelae ($p = 0.007$) at a mean follow-up of 15 months . In contrast to the study performed in Dallas with cefuroxime and dexamethasone, no difference in the incidence of hearing loss due to bacterial meningitis was noted between the two study groups both treated with cefotaxime with and without dexamethasone in Costa Rica. Questions remain regarding the efficacy of dexamethasone in the management of pneumococcal meningitis since the majority of children in both of these studies had meningitis due to Hib. One study conducted in Egypt concluded that adjunctive dexamethasone therapy was associated with decreased mortality (13.5 vs. 40.7%) and hearing loss (0 vs. 12.5%) compared with placebo treatment in children and adults with pneumococcal meningitis. The reader should be familiar with the statement of the Red Book Committee regarding the use of dexamethasone in bacterial meningitis and be aware of the dose, duration and possible complications of dexamethasone as well as other indications to discontinue its administration. The efficacy and safety of dexamethasone in neonatal meningitis are currently under investigation.

Long-Term Prognosis

Focal neurologic deficits such as hemiparesis are related to vascular insufficiency due to vasculitis and/or thrombosis and occur in approximately 12% of children with bacterial meningitis. In follow-up studies these deficits generally resolve but persistent deficits are associated with cerebral defects demonstrated by CT scan or other techniques. Sensorineural hearing loss is caused by labyrinthitis and inflammation of the auditory nerve and is usually permanent. Routine hearing assessment prior to or shortly after discharge is recommended. Full otologic and audiologic evaluations are indicated if hearing losses are detected. Occasionally, a child with normal hearing results at discharge may be found to have an important hearing loss several months after discharge. Alternatively hearing may improve with time. The mechanisms by which steroids may alter the development of sensorineural deafness are unknown. Presumably, there is less inner ear inflammation and subsequent damage to the auditory nerve or structures within the cochlea.

Learning disabilities, behavior disorders, and other subtle defects in intellect may result from bacterial meningitis. For children who do not have obvious serious brain injury, the follow-up studies of Hib meningitis from Pittsburgh provide reasons to be cautiously optimistic regarding the ultimate outcome of this infection. The 22 patients studied scored close to the predicted means for academic achievement tests in reading, spelling, and arithmetic. The major finding was that 13 patients, but only 1 of 11 control siblings, received additional family help with homework. Thus, it appears that most children following Hib meningitis can be expected to perform normally in school, although they may require more help outside the classroom than other children.

Long-term follow-up for children with meningitis due to group B streptococcus is also available. Major neurologic sequelae are more common in this group compared to children followed for Hib meningitis. Wald et al. compared 34 children who had group B streptococcus meningitis to 21 control siblings. The two groups did not differ significantly on psychometric testings except for the Beery Test of Visual-Motor Integration.

Herpes simplex Meningoencephalitis

Herpes simplex encephalitis (HSE) is the only commonly occurring viral meningoencephalitis for which virus-specific treatment is available. Both vidarabine and acyclovir have been proven efficacious in the treatment of HSE. In a randomized study involving 69 patients over 6 months of age with biopsy-proven disease, acyclovir-treated patients had lower morbidity and mortality and a higher percent of patients functioning normally (38%) than patients treated with vidarabine. In this study patients with Glasgow coma score <6 had a poor outcome. As a result of this study acyclovir (30 mg/kg/day in three divided doses intravenously) is considered the treatment of choice for HSE.

A major problem that remains is the timely diagnosis of HSE. Isolation of herpes simplex from brain tissue is the gold standard for documenting HSE. Obviously this is a procedure reluctantly recommended, especially since acyclovir can be administered without much difficulty. However, only 35% of patients over 6 months of age with suspected HSE actually had biopsy proven HSE in the collaborative study. In some instances other treatable infections or noninfectious conditions such as toxoplasmosis, cryptococcal infection, or *Mycobacterium tuberculosis* are discovered by brain biopsy. Soong et al. developed a logistic regression model to predict the probability of HSE based on certain clinical variables including the presence of a focal brain scan, focal computed tomography, CSF WBC > 5/mm, age > 39 years, and focal electroencephalogram. Obviously not all variables will be available in a timely manner and some tests may be difficult to obtain at any time. Furthermore, the validity of this model in children is unclear since less than 25% of the population studied were less than 10 years old. In addition, the changes observed by CT scan do not occur early in the disease process and may not be observed until the second week of illness.

Magnetic resonance imaging (MRI) appears to detect changes consistent with HSE earlier than CT scan. In comparative studies MRI demonstrated the typical unilateral or bilateral inflammation of the temporal lobes several days before changes on CT scan were evident. Thus MRI is the neuroimaging procedure of choice, if HSE is seriously suspected.

Acyclovir and vidarabine have also been compared in neonatal herpes simplex infection in a multicenter trial. The mortality rate for encephalitis was 14% for both the 36 babies receiving vidarabine and the 35 babies given acyclovir. At 1 year of age 43% of vidarabine-treated and 29% of acyclovir-treated infants were developing normally. The investigators concluded that acyclovir and vidarabine were equivalent in the treatment of neonatal herpes simplex virus infection. Since acyclovir is more convenient to administer on an every 8-hour schedule versus a continuous 12-hour infusion for vidarabine, acyclovir is considered the agent of choice for this infection. The optimal duration of acyclovir therapy for neonatal HSE is not clear. In the multicenter study therapy both drugs were continued for 10 days. Seven of 87 surviving babies (8%) had a recurrence of disease, which included CNS infection, within one month of completing therapy (6 vidarabine, 1 acyclovir). Prolongation of therapy to 21 days has been suggested as a possible approach to prevent these relapses, but this remains speculative.

Brain Abscess

The most recent reviews of brain abscesses in children indicated that ear, nose and throat infections such as chronic otitis media and mastoiditis or sinusitis, as well as cyanotic congenital heart disease remain the most common predisposing factors for brain abscess formation in children. Alpha-streptococci, and microaerophilic or anaerobic streptococci are associated with cyanotic congenital heart disease whereas *Staphylococcus aureus* is most commonly isolated when head trauma or neurosurgery has predisposed to abscess formation. Circumstantial evidence indicates isolates of *Streptococcus milleri* have a propensity to cause abscesses, including within the central nervous system. A polysaccharide capsule is one virulence factor possibly related to the abscess-creating ability of the *S. milleri* group.

Brain abscess formation as a complication of bacterial meningitis is rare in infants and children, but is seen more commonly in neonatal meningitis. *Citrobacter diversus,* in particular, is associated with brain abscess formation in 75% of neonates with meningitis due to this organism. A minor outer membrane protein (32 kD) is associated with 90% of *C. diversus* strains causing meningitis, but in only 10% of strains isolated from non-CSF sources.

Animal models have been useful for studying the pathogenesis of brain abscesses. In the canine model four stages of the evolution of a brain abscess were observed: early cerebritis (day 1-3), late cerebritis (day 4-9), early capsule formation (day 10-13), and late capsule formation (day > 14).

Animal models of brain abscess usually involve direct inoculation of bacteria into brain tissue with or without concomitant cerebral injury or placement of foreign material. These models may be appropriate for studying abscesses due to contiguous structures or head trauma. In a neonatal rat model, *C. diversus* was capable of causing brain abscesses following intraperitoneal inoculation and hematogenous spread to the brain. In this model it appeared that the *C. diversus* penetrated through the ependymal lining to cause periventricular abscesses.

The clinical manifestations of brain abscesses are nonspecific. The physician must be alert that persistent headache, fever, and symptoms of increased intracranial pressure in a child with predisposing factors raise the possibility of the presence of a brain abscess. Computed tomography remains an important neuroimaging technique for the diagnosis of brain abscesses in children. The location, size, number, presence of surrounding edema and capsule formation, as well as evidence of mass effect, are generally easily determined by CT scan. MRI does have some advantages over CT scan for the evaluation of brain abscesses, including lack of radiation exposure and greater detail of the abscess cavity and surrounding tissues. In addition, enhancement by gadolinium-DTPA during MRI appears to be superior to contrast studies by CT scan.

It is optimal to obtain brain abscess material for culture and antimicrobial susceptibility testing of isolates. Aspiration of the brain abscess can be accomplished with little risk when performed under CT guidance. This is especially true for lesions that are superficially located. Antimicrobial therapy initially is administered empirically based on the most likely pathogens associated with the predisposing factors of the host, but subsequently is directed against the pathogens isolated. Penicillin plus chloramphenicol is a reasonable combination for empiric therapy of brain abscesses associated with otitis media, mastoiditis and sinusitis. Recent experience with metronidazole indicates that it may be a more desirable agent than chloramphenicol for this infection. Metronidazole is bactericidal for most anaerobes and penetrates brain tissue well. In contrast to chloramphenicol, metronidazole does not require monitoring of serum levels and adverse effects are unusual but do include CNS toxicities such as headache, drowsiness and vertigo. The duration of therapy is generally 6-8 weeks, but there are no studies specifically addressing this issue.

The optimal surgical approach to management of brain abscesses is unclear. Currently one or more aspirations of the brain abscess in conjunction with prolonged intravenous antibiotics is at least comparable, if not superior, to excision of the abscess. Certainly the approach to each patient is an individual one which considers the age of the patient, underlying disease, number and location of the abscesses as well as the experience of the neurosurgeon.

The progress of therapy in patients with a brain abscess is monitored by both clinical response and changes observed on neuroimaging. Again, MRI is favored over CT scan due to no exposure to radiation despite repeated imaging; MRI may also show evidence of decreasing inflammation more reliably than CT scan. Duration of antibiotic therapy should not be based on the CT or MRI findings returning to normal since contrast enhancement may be evident for months despite adequate therapy.

FUNGAL INFECTIONS

Cryptococcal Meningitis

Cryptococcus neoformans is the most common fungus causing meningitis in patients with AIDS, although this remains a relatively rare infection in pediatric AIDS patients. The combination of amphotericin B plus flucytosine is the recommended treatment for cryptococcal meningitis. If the patient has no underlying disease, no neurologic complications and is not receiving immunosuppressive therapy, a 4-week course of treatment is satisfactory. If these risk factors are present, a minimum of 6 weeks of combined therapy is standard. In patients with AIDS, antifungal therapy is continued indefinitely due to an over 50% relapse rate once therapy is completed.

Since amphotericin B requires parenteral administration and is associated with adverse effects such as chills, fever, and reduced renal function, alternative therapies are actively being investigated. Fluconazole is an oral triazole antifungal agent that penetrates well into CSF with a long half-life within the CNS compartment. Fluconazole may be useful to prevent relapses of cryptococcal meningitis following a full course of therapy with amphotericin B plus flucytosine in AIDS patients, but should not be used as primary therapy. Experience in pediatric patients with fluconazole is limited, although fluconazole maintenance therapy in adolescents could be considered based on the adult trials.

Histoplasmosis

As with *C. neoformans* new knowledge regarding CNS infections due to *Histoplasma capsulatum* focuses on patients with AIDS. New radioimmunoassays and enzyme-linked immunosorbent assays (ELISA) have been developed to detect the polysaccharide antigen of *H. capsulatum*. As with other polysaccharide antigens, *H. capsulatum* antigen is most readily found in urine compared to serum or CSF in

TABLE 15.1. NEUROLOGIC FINDINGS IN 104 PATIENTS WITH CENTRAL NERVOUS SYSTEM HISTOPLASMOSIS

Finding	*Percentage*
Depressed consciousness	29
Headache	24
Confusion	22
Cranial nerve deficit	19
Other focal deficit	16
Seizure	14
Personality change	12
Ataxia	11
Papilledema	3

patients with meningitis. Antibody to *H. capsulatum* in CSF also is helpful in establishing this diagnosis. In a review of *H. capsulatum* infections of the CNS, Wheat et al. tabulated the neurologic findings in 104 patients. The number of children included in the review was not stated. The majority of the patients were symptomatic for over 2 months. Meningitis was present in 60% of cases and cerebral mass lesions were present in almost 40%. *H. capsulatum* should be considered in the differential diagnosis of chronic meningitis in a child, especially if underlying immunodeficiency or immunosuppression is present. The optimal treatment of CNS histoplasmosis is not known. Wheat and co-workers recommend at least a 35 mg/kg course of amphotericin B. Patients with AIDS may benefit from suppressive therapy to prevent relapses which are common. The role of intraventricular administration of amphotericin B is unclear.

HELMINTHIC INFECTIONS

Neurocysticercosis

Cysticercosis is the most common helminth infection of the CNS in children and is due to *Taenia solium,* a common intestinal tapeworm worldwide. Neurocysticercosis is most frequent in developing areas of the world such as Mexico and Central and South America. The true prevalence of this infection is unknown, but in autopsy studies up to 3.8% of the general population in endemic regions have evidence of infection. In the United States the typical child has traveled to or grew up in an endemic region. The most common symptoms in children are seizures which may be focal or generalized and are related to parenchymal invasion, subsequent cyst formation and the ensuing inflammatory response of the host. CT scan and MRI are both useful for detecting intracranial abnormalities suggestive of neurocysticercosis. CT scan is best for depicting intraparenchymal calcifications but MRI is superior for visualizing intraventricular cysts.

Mitchell and Crawford reviewed the experience with presumed intraparenchymal cerebral cysticercosis at the Children's Hospital of Los Angeles over a 6-year period. Fifty-two patients when first seen were between 21 months and 20 years of age. Fifty children had seizures; 42 children had an acute inflamed lesion with or without concomitant calcified granulomas. Of the children with acute lesions (lesions with surrounding edema or contrast enhancement), 70% had no seizures after their initial presentation. Twenty-two percent had seizures while receiving anticonvulsants or after these medications were discontinued.

Cerebrospinal fluid was examined in 56% of the patients and only 5 children had a WBC count > 5/mm. Peripheral eosinophilia was present in 2 children both of whom also had intestinal parasites. Thirty-two of 41 children with acute and 6 of 9 with inactive disease had a single lesion on CT scan. None of these patients received antiparasitic therapy including praziquantel and the authors concluded that praziquantel is not necessary for pediatric neurocysticercosis when a single inflamed intraparenchymal cyst is present.

Other experts recommend antiparasitic therapy for patients with viable parenchymal cysts. Praziquantel is an isoquinoline which has activity against *Taenia solium.* In adult studies praziquantel therapy was associated with a reduction in the number and size of parenchymal lesions in a selected group of patients. Similar studies are not available in children. If praziquantel is administered some patients may show increased neurologic symptoms (headache, nausea, seizures) presumably due to exacerbation of the inflammatory response by the release of antigens by the dying cyst. Whether routine steroid administration prior to praziquantel therapy is always necessary to avoid this response is unknown, however, plasma levels of praziquantel are reduced by - 50% when dexamethasone is administered concomitantly. Albendazole is another agent evaluated in Mexico for neurocysticercosis. In adults albendazole and praziquantel appear equivalent in efficacy in a small number of patients which did not include children. If treatment is instituted repeat neuroimaging is best done 3 months after the completion of therapy.

Protozoan Infections

The serologic diagnosis of congenital toxoplasmosis may be improved by an ELISA for IgA antibodies, although at the present time this test is not widely available.

In France the prenatal diagnosis of congenital toxoplasmosis can be established by monitoring maternal serology and culture and serology of fetal blood and amniotic fluid. Spiramycin treatment of pregnant women in whom this diagnosis is established reduces the clinical severity of the congenital toxoplasmosis in the infant. The optimal treatment of congenital toxoplasmosis is unknown. Presently combinations of pyrimethamine, trisulfpyrimidine and spiramycin are being evaluated in a randomized prospective study. The focus of research over the past several years for *Toxoplasma gondii* CNS infections has been in AIDS patients.

Tuberculous Meningitis

The rapid diagnosis of tuberculous meningitis is an area of great interest to investigators. Assays using CSF under development for this infection include adenosine deaminase activity, enzyme-linked immunoassay of mycobacterial antigens or antibody, and polymerase chain reaction (PCR). Unfortunately, there is no widely available and accepted laboratory technique to rapidly diagnose Tb meningitis.

The recommended treatment of tuberculous meningitis in children has recently been revised by the Committee on Infectious Diseases of the American Academy of Pediatrics in collaboration with the Centers for Disease Control and the American Thoracic Society. A 12-month course of therapy is recommended. The patient is initially treated with daily isoniazid (10-15 mg/kg/day), rifampin (10-20 mg/kg/day), pyrazinamide (20-40 mg/kg/day) and streptomycin (20-40 mg/kg/day i.m.) for 2 months (kanamycin may be used when streptomycin cannot be obtained). Isoniazid and rifampin are then administered daily (or twice weekly under direct observation) for 10 months. A 6-month regimen containing pyrazinamide may also prove efficacious but requires further study. The use of prednisone or dexamethasone for 4-6 weeks followed by tapering doses over 2-3 weeks is also considered beneficial 55]. Mortality and neurologic sequelae of Tb meningitis were reduced by adjunctive dexamethasone in a randomized study conducted in Egypt.

16

VIBRIO

Most clinical laboratories isolate many strains of *Enterobacteriaceae* each year, and workers feel secure with the family. Unfortunately, this is not the case with the genus *Vibrio,* which was once relegated to the category "strangers in the world of clinical microbiology." *Vibrio* species are probably quite rare in many inland laboratories in the United States, where contact with the ocean or seafood is rare. Laboratories which serve coastal areas are more likely to encounter *Vibrio* strains. This situation in the United States is in contrast to some undeveloped areas of the world where cholera and its causative organism, *Vibrio cholerae,* are very common. *Vibrio* cultures are also encountered in countries where there are cultural differences in food preference. For example, in Japan, where fish and other seafoods are eaten raw, *Vibrio parahaemolyticus is* one of the most common causes of gastroenteritis. In this chapter we will discuss methods which will apply to this diverse spectrum of *Vibrio* distribution. The material which applies to *Vibrio* species as a group is discussed first and is followed by a discussion of each of the 10 *Vibrio* species which occur in clinical specimens.

Because of space limitations, many primary references cannot be given. However, several excellent reviews are available on the family *Vibrionaceae,* the marine vibrios, the genus *Vibrio, V. cholerae, V. parahaemolyticus,* the vibrios not associated with human disease, and

Vibrio toxins. For the other species, there is much less literature, and we will give some primary references.

NOMENCLATURE AND CLASSIFICATION

In the last few years the genus *Vibrio* has changed from a poorly characterized heterogeneous group of organisms to a wellunderstood natural group. This has been due to the removal of "nonfermentative vibrios," "anaerobic vibrios," and "microaerophilic vibrios" to other genera such as *Campylobacter (Vibrio fetus), Wolinella (Vibrio succinogenes), Pseudomonas,* and *Alteromonas*. The genus *Vibrio* is now much more homogeneous. The species currently recognized as belonging to genus *Vibrio* are listed in. The genus *Vibrio* is classified in the family *Vibrionaceae* along with three other genera, i.e., *Photobacterium, Aeromonas,* and *Plesiomonas.* gives the guanine-plus-cytosine ratios of the genera of *Vibrionaceae,* which can be used as a first step in determining the relatedness of genera. The genus *Vibrio is* closely related to the genus *Photobacterium,* which would be expected because of their phenotypic similarities, but it is more distantly related to the genus *Aeromonas* and to the family *Enterobacteriaceae*. More distant relatives include the purple photosynthetic bacterium genus *Chromatium* and the nonfermentative genera *Pseudomonas and Acinetobacter*. Most of the *Vibrio* species are not closely related to each other in a phylogenetic sense; thus *Vibrio is* a heterogeneous genus, similar to the genus *Pseudomonas*. The genus *Vibrio is* maintained as a matter of convenience, since most *Vibrio* strains share common phenotypic properties.

Vibrio Classification for the Clinical Laboratory

The identification of *Vibrio-Photobacterium* species can be greatly simplified from the 31 species given in Table 1 if we consider only the species which cause human disease or occur in human clinical specimens. This simplified classification for the genus *Vibrio* recognizes the 10 named species and considers all the other named species of *Vibrio* and *Photobacterium* as "marine vibrio species" (synonym, marine vibrios).

We define a "marine vibrio" as *a Vibrio* or *Photobacterium* strain which is oxidase positive, ferments Dglucose, does not grow in nutrient broth without added NaCl, but does grow in nutrient broth with added NaCl or "sea salts." Its properties do not fit any of the 10 "clinical species" of the genus *Vibrio*. Many of these marine vibrios require 1 to 2% NaCl for growth and grow better at a lower temperature than at 36°C. Some are bioluminescent.

Many of the organisms isolated from ocean or estuary water or from the plants and animals which live in this water will belong to the marine vibrio group. These organisms can be identified to species, but this task is much more difficult because the identification methods are very different from those used in clinical laboratories.

Reporting Vibrio Species

Based on the above discussion, we recommend that clinical laboratories report the 10 *Vibrio* species given in the top of Table 1 by genus and species name and report the other species as "marine vibrios." Reference laboratories may also wish to adopt this simplified classification until all the named *Vibrio* and *Photobacterium* species have been characterized by the same methods and an identification matrix becomes available. We are collecting these data for a data matrix similar to the one for all the species of *Enterobacteriaceae*.

CLINICAL SIGNIFICANCE

Vibrio species cause a number of human infections which can be classified as intestinal or extraintestinal. Each of these infections is described briefly below but is discussed in more detail in the section for each species.

TABLE 16.1. CLASSIFICATION OF THE GENUS *VIBRIO* AND SOME RELATED ORGANISMS

Vibrio species which occur in human clinical specimens		
V. alginolyticus	*V. hollisae*	
V. cholerae	*V. metschnikovii*	
V. damsela	*V. mimicus*	
V. fluvialis	*V parahaemolyticus*	
V. furnissii	*V. vulnificus*	
Vibrio and *Photobacterium* species which do not occur in human clinical specimens		
V. aestuarianus	*V natriegens*	*P. angustum*
V anguillarum	*V nereis*	*P. leiognathi*
V. campbellii	*V. nigripulchritudo*	*P. phosphoreum*
V. costicola	*V. ordalii*	
V. diazotrophicus	*V orientalis*	
V. fischeri	*V. pelagius*	
V. gazogenes	*V. proteolyticus*	
V. harveyi	*V. splendidus*	
V. logei V tubiashii		

Infections of the Intestinal Tract

Five *Vibrio* species have been shown to cause or be associated with diarrhea. *V. cholerae is* well known as the cause of cholera which is a distinct clinical entity in its most severe form but in its milder form may be hard to distinguish from other secretory diarrheas. *V. parahaemolyticus is* a well-documented cause of acute gastroenteritis. More recently, *Vibrio fluvialis, V. hollisae,* and *V. mimicus* have been implicated as causes of diarrhea. *Vibrio furnissii* has been isolated from a few cases of diarrhea, but there is no real evidence that it can actually cause diarrhea.

TABLE 16.2. SOME CHARACTERISTICS OF THE FAMILIES *ENTEROBACTERIACEAE, VIBRIONACEAE,* AND PSEUDOMONADACEAE

	Property of family (typical genus)		
Characteristic	*Enterobacteriaceae (Escherichia)*	*Vibrionaceae (Vibrio)*	*Pseudomonadaceae (Pseudomonas)*
Grow aerobically	+	+	+
Grow anaerobically	+	+	—
Ferment D-glucose	+	+	—
Flagella locations	Peritrichous	Polar	Polar
Oxidase	—	+	+
Often bioluminescent	—	+	—
Na^+ required by many species	—	+	—
Inhibited by vibriostatic compound 0129	—	+	—

Extraintestinal Infections

Vibrio strains have been isolated frequently from certain extraintestinal infections but are rarely isolated from others. Vibrios are often isolated from blood, wounds of arms and legs, infected eyes and ears, and the gallbladder. They are rarely reported from patients with meningitis, pneumonia, and infection of the reproductive organs or urinary tract, but they can occasionally occur. Within the genus *Vibrio* there is a division between the species which cause intestinal infection and those which cause extraintestinal infections; however, this division is not absolute. The pandemic strain of *V. cholerae* (serogroup 01, cholera toxin') seems to be well adapted to the human intestinal tract; seldom is it found at other sites. Similarly, many of the wound infections and septicemia reportedly due to *V. parahaemolyticus* were really due to *Vibrio vulnificus,* which was not described until recently.

TABLE 16.3. PROPERTIES OF THE FOUR GENERA OF THE FAMILY *VIBRIONACEAE*

Property	*Property of genus*			
	Vibrio	*Photobacterium*	*Aeronvonas*	*Plesiomonas*
Guanine + cytosine content of DNA (mol%)	38-51	40-44	57-63	51
Associated with human infections	+	—	+	+
Na^+ required for growth (or stimulates growth)	+	+	—	—
Sensitive to vibriostatic compound 0129	+	+	—	+
Lipase	+	V	+	—
n-Mannitol fermentation	+	—	+	—
Sheathed polar flagella	+	—	—	—
Accumulate poly-β-hydroxy butyrate,	—	+		

V. vulnificus is well established as an important cause of septicemia (often fatal) and wound infections. *Vibrio damsela* is associated with human wound infections), but its causal role needs strengthening. *Vibrio alginolyticus* has been isolated from several types of soft-tissue infections. *Vibrio metschnikovii is* usually an environmental organism, but it has been isolated in a single case of peritonitis in a patient with an inflamed gallbladder. Most, but not all, of these infections are associated with seawater contact.

USUAL LABORATORY TECHNIQUES

Many of the techniques used in clinical microbiology and enteric bacteriology laboratories work well with the genus *Vibrio*. However, there are some specialized items that are recommended for laboratories which often isolate and identify *Vibrio* species and for clinical laboratories that want to increase their capability. Two factors can complicate the isolation and identification *of Vibrio* species. Sometimes *Vibrio* cultures will not grow well on the highly selective media used to isolate enteric pathogens. The other factor is that the halophilic species of *Vibrio* need added NaCl for optimum growth and activity, and several laboratory media have suboptimum amounts of Na+ (less than *0.5%* NaCI). Solutions to these problems are discussed below. Four approaches to the isolation of *Vibrio* species are as follows.

1. Use normal procedures, and make no special effort to search for *Vibrio* species.

2. Use normal procedures and plating media, and look for oxidase-positive colonies.

3. Incorporate TCBS agar as an extra plate for stool cultures and also for other likely specimens such as those from wounds, blood, eye, and ear.

4. Use other special procedures to enhance the isolation of V *cholerae, V. parahaernolyticus,* and other *Vibrio* species.

The approach adopted by a particular laboratory will probably depend on the frequency with which *Vibrio* cultures are encountered. For example, a small laboratory in the American Midwest will rarely see a *Vibrio* culture, and approach 1 or 2 will probably be adequate. A large clinical laboratory near the ocean may wish to use TCBS agar routinely as a plating medium. There are both advantages and disadvantages to the routine use of TCBS medium. On the positive side, this medium brings immediate attention to a possible *Vibrio* isolate, whose colonies may not be obvious onother plating media. On the negative side, the routine use of TCBS may not be cost-increased dramatically (to *15%).* effective. In an 18-month study, one of us (M.T.K.) found that every *Vibrio* isolate found on TCBS medium was also detected on the other plating media used (which were screened for oxidasepositive colonies). Other negative factors include the facts that different lots of commercial TCBS medium vary in their selectivity and that some *Vibrio* species or strains do not grow well on TCBS . These positive and negative aspects of the routine use of TCBS medium must be considered.

TABLE 16.4. GROWTH OF *VIBRIO* CULTURESON TCBS MEDIUM

Organism	*Colony appearance on TCBS' (%):*		*Growth-plating efficiency*
	Green	*Yellow*	
V cholerae	< 1	>99	Good
V. rnimicus	100	0	Good
V rnetschnikovii	0	100	May be reduced
V. hollisae	100	0	Very poor
V. damsela	95	5	Reduced at 36⁰C
V. fluvialis	0	100	Good
V furnissii	0	100	Good
V alginolvticus	0	100	Good
V parahaemolvticus	99	1	Good
V vulnificus	90[1]	10	Good
Marine vibrios	Variable	Variable	Variable

An active laboratory on the Indian subcontinent, where cholera is epidemic or endemic, will probably want to look specifically for *V. cholerae. A* laboratory interested in the environmental aspects of the genus *Vibrio may* want to use the specialized procedures in approach 4 to increase isolation rate.

COLLECTION, TRANSPORT, AND STORAGE OF SPECIMENS

Extraintestinal

The usual procedure for the collection and processing of an extraintestinal specimen is followed. There are no special procedures for *Vibrio* species.

Feces Collection

Stool specimens should be collected early, preferably within the first 24 h of illness, and before the patient has received any antimicrobial agents. Fluid stool may be collected by inserting a petrolatumlubricated soft-rubber catheter into the rectum. Rectal swabs may also be used. Although highly efficient in the acute phase of illness, rectal swabs are probably less satisfactory for convalescent patients or transiently infected asymptomatic persons. The administration of purgatives has been reported to increase the detection of persons excreting small numbers of vibrios. Vomitus, if available, may also be collected for culture.

Transport

Whenever possible, stool or rectal swab specimens should be inoculated onto isolation plates with minimal delay. The viability of *Vibrio* species is well maintained at the alkaline pH of ricewater stool but is unpredictable in formed stools. Vibrios are very susceptible to desiccation; hence, specimens must not be allowed to dry.

When there will be a delay in plating a culture (especially when it must be transported by courier), rectal swabs or fecal material should be placed in the semisolid transport medium of Cary and Blair, which maintains the viability of *Vibrio* cultures for up to 4 weeks. Buffered glycerol-saline, often used in enteric bacteriology, is an unsatisfactory transport medium even for short periods. Tellurite-taurocholate-peptone broth has been extensively used with success as an enrichment transport medium at the Cholera Research Laboratory in Dacca, Bangladesh, where specimens collected in the field are generally plated within 12 to 24 h. In the absence of available suitable transport media, strips of blotting paper may be soaked in liquid stool and inserted into airtight plastic bags. Specimens collected in this way may remain viable for up to 5 weeks.

Storage

Specimens in transport medium may be shipped to the laboratory without refrigeration.

Direct Examination

Direct examination of stool material is not recommended for general purposes. However, stools may be examined by dark-field microscopy for the characteristic size, shape, and darting motility of vibrios, especially after a brief incubation in broth.

CULTURE AND ISOLATION

A simple procedure will probably result in a higher isolation rate for *Vibrio* species. Hemolytic colonies on sheep blood agar plates are tested for their oxidase reaction. This procedure will detect strongly hemolytic colonies of *Aeromonas,* some *Vibrio* species, and also some weakly hemolytic *Vibrio* and *Plesiomonas* cultures. Oxidase testing can also be done on nonhemolytic colonies.

Extraintestinal Specimens

Extraintestinal specimens are usually processed with no particular attention to *Vibrio*. The *Vibrio* species of medical importance grow well on blood agar and may also grow on MacConkey agar. However, a more thorough search for *Vibrio* isolates can be done with oxidase testing or by including a plate of TCBS medium.

Stool Specimens

Stool specimens will usually be plated onto a nonselective medium such as blood agar; onto selective media such as MacConkey agar, phenylethyl alcohol-blood agar, xylose-lysine-deoxycholate agar, and Hektoen enteric agar; and into gram-negative broth (Hajna). Specialized plating media for *Campylobacter, Yersinia,* and *Salmonella* species are used in some laboratories. *Vibrio cultures* will grow well on blood agar, where they will be betahemolytic (*V. cholerae* non 01 and *V. cholerae 01* of the eltor biotype), alpha-hemolytic (*V. vulnificus* and many others), or nonhemolytic. *Vibrio* cultures usually grow on MacConkey agar (but sometimes with a reduced plating efficiency) and will appear as colorless (lactose-negative) colonies. Oxidase testing can be done on colonies grown on blood agar and usuallyy on lactose-negative colonies on selective media; however, lactose-positive colonies from selective media can give a false-negative oxidase reaction. *Vibrio* cultures often do not grow well on the more selective enteric plating media.

Oxidase Testing of Colonies on Primary Plates

Many laboratories have adopted oxidase testing of colonies on primary plates as an alternative to adding selective media or special procedures for *Vibrio* species. All the *Vibrio* species which occur in clinical specimens are strongly oxidase positive in contrast to species in the family *Enterobacteriaceae,* which are oxidase negative. The lone exception is the oxidasenegative species *V. metschnikovii,* which is extremely rare in clinical specimens. This difference in oxidase reaction makes it easy to spot *Vibrio* colonies among colonies of *Enterobacteriaceae. Two* approaches have been used to spot *Vibrio* cultures this way. (The oxidasepositive species *Pseudomonas aeruginosa* would cause many false positives, but it can usually be recognized by, its typical odor and colony morphology on plating media.)

Method 1. Five to 10 colonies from the blood agar are individually touched and spot tested for their oxidase reaction by Kovac's method.

Method 2. A drop of Kovac's oxidase reagent is added to an area of the blood agar plate where the colonies are crowded but still separated. The drop should cover 50 to 100 colonies. An alternative is to pick growth from a crowded area and do the Kovac's oxidase test. If oxidase-positive colonies are present, individual isolated colonies (not exposed to the reagent) from an area of less crowding are tested for their oxidase reactions as described above. Since most specimens will have no oxidase-positive colonies (other than *P. aeruginosa),* this method reduces the number of colonies which require individual screening. Oxidasepositive colonies (other than *P. aeruginosa)* detected by the above method are then identified to species. This method has another advantage because it also detects cultures *of Aeromonas* and *Plesiomonas,* which are now being looked for in diarrheal stools. In many areas of the United States this method has proved successful but not cost-effective, because of the low yield of *Vibrio, Plesiomonas,* and *Aeromonas* strains.

METHODS FOR *VIBRIO* ISOLATION

Isolation from Feces

This method should also be useful in the isolation of other *Vibrio* cultures from feces and other specimens. TCBS agar and enrichment in alkaline peptone water select for *Vibrio* species. These special procedures are integrated into the normal laboratory procedures, but they are used in addition to the normal isolation media.

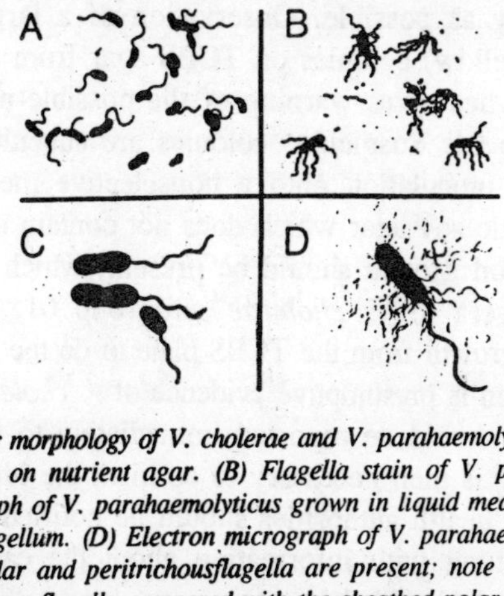

Figue 4.2: Cellular morphology of V. cholerae and V. parahaemolyticus. (A) V. cholerae from 18-h culture on nutrient agar. (B) Flagella stain of V. parahaemolyticus. (C) Electron micrograph of V. parahaemolyticus grown in liquid medium. Note the single, sheathed polar flagellum. (D) Electron micrograph of V. parahaemolyticus grown on a solid medium: Polar and peritrichousflagella are present; note the different size and shape of peritrichous flagella compared with the sheathed polar flagellum.

Microscopic Examination of Growth

In 1883, Koch noted the appearance of small curved rods in the rice-water stool of cholera patients. These typical curved rods were not seen in the feces of patients without cholera. A number of other authors have made the same observation and commented on the possible usefulness of this observation. *V. cholerae* has considerable variability in its cellular morphology, which includes typical curved rods, straight rods, short noncurved rods, and involution forms). Often these forms can be seen in the same culture. For this reason, the microscopic examination of feces or cultures does not have a prominent role today. However, the finding of typical curved forms can be used as presumptive evidence for the presence of *Vibrio* strains. *Vibrio* cultures grown in liquid media have polar flagella, but many strains have peritrichous flagella when grown on solid media. Flagella stains have no place in the routine identification of vibrios.

IDENTIFICATION

Because pathogenicity and clinical significance vary among different species, a *Vibrio* strain should be identified to species. There are many different approaches to identification in general . One important question needs an immediate answer. Is the *Vibriolike* organism isolated from a case of choleralike diarrhea the pandemic strain of *V. cholerae?*

Ruling out *V. cholerae* serogroup 01

It is very important to give a definitive answer to the above question as quickly as possible. Observation of a large number of sucrose-positive (yellow) colonies on TCBS agar from a patient with severe watery diarrhea gives warning of the possible presence of *V. cholerae* serogroup 01. Suspicious colonies are subcultured early in the day by heavy inoculation onto a nonselective medium such as blood agar or tryptic soy agar which does not contain carbohydrates. After 5 to 8 h, good growth should be present, which can be tested in polyvalent antisera to *V. cholerae* serogroup 01. (Experienced workers often use growth from the TCBS plate to do the agglutination.) Positive agglutination is presumptive evidence of *V. cholerae* serogroup 01, and the result should be reported immediately to the physician. Biochemical testing is then necessary to confirm the identification as *V. cholerae*. Local health authorities should be notified immediately, and the culture, along with information about the case, should be forwarded to the state health laboratory.

Biochemical Identification

Most laboratories use media and tests designed to identify cultures of the family *Enterobacteriaceae*. These procedures work well for *V. cholerae* and *V. mimicus* because these species have only a slight requirement for Na^+, which is fulfilled by the amount of NaCl in the medium. However, most of the halophilic *Vibrio* species require much more Na^+ for growth and expression of their metabolic pathways. Some of the media for doing biochemical tests do not contain enough NaCl for these halophilic species. The standard enteric test is listed first, and the percentage of positive responses for each species is given. The percentage of positive responses for each species is then given for a modified medium with 1% NaCl added. More *Vibrio* species are indole positive when the medium is changed from peptone water to heart infusion broth, and more strains are VogesProskauer positive when the reagent for detecting acetyl methl carbinol contains alphanaphthol. The vast majority of *Vibrio* cultures isolated from clinical specimens will be easily identified as one of the species listed. The six tests which we find very helpful in dividing the 10 species into 5 groups are: requirement for Na^+, oxidase, nitrate reduction to nitrite, arginine dihydrolase, lysine decarboxylase, and ornithine decarboxylase.

Growth in the absence of added Na^+ (growth in nutrient broth . with 1% NaCl, but no growth in nutrient broth) is the essential test for the differentiation of *V. cholerae* and *V. mimicus* from the other eight *Vibrio* species and from the marine vibrios. Table gives the key characteristics for *V. cholerae* and *V. mimicus* and the key tests for differentiating these two species, which are very similar biochemically. The tests which are used to define the classical and eltor biogroups of *V. cholerae*, *V. metschnikovii is* easily differentiated because it is oxidase negative and nitrate negative. All of the other vibrios are positive for these two tests. *V. hollisae is* negative for arginine dihydrolase and for lysine and ornithine decarboxylases (triple decarboxylase negative), which differentiates it from the remaining *Vibrio* species. The remaining six species are subdivided into two groups, i.e., the group which is arginine dihydrolase positive and includes *V. darnsela, V. fluvialis,* and *V. furnissii* and the group which is argininc dihydrolase negative but lysine decarboxylase positive and includes *V. alginolvticus, V. parahaernolvticus,* and *V. vulnificus*.

Commercial identification Systems

Many laboratories use commercial kits to identify clinical isolates.

TABLE 16.5. THE EFFECT OF MEDIA (ESPECIALLY NA^+ CONTENT) AND TEST CONDITIONS ON BIOCHEMICAL TEST RESULTS

Test	Positive for Vibrio species:									
	V. chol-erae	*V mime-cus*	*V. metschni-kovii*	*V. holli-sae*	*V. dam-sela*	*V. fluvi-alis*	*V. fur-nissii*	*V. algino-lyticus*	*V. P. haemo-lyiticus*	*V. vul nificus*
Indole production										
Peptone water	86	94	22	30	0	0	0	14	35	39
Peptone water + 1% NaCI	86	98	26	93	0	0	0	24	80	65
Heart infusion + 1% NaC1	97	95	17	97	0	13	11	42	89	94
Methyl red test										
Standard	25	14	17	NG	NG	0	0	NG	NG	NG
1% NaC1	99	99	96	0	100	96	100	77	78	79
Voges-Proskauer										
Standard	74	2	26	NG	NG	0	0	NG	NG	NG

Table Contd.

Table Contd.

Test	Positive for Vibrio species:									
	V. chol-erae	***V mime-cus***	***V. metschni-kovii***	***V. holli-sae***	***V. dam-sela***	***V. fluvi-alis***	***V. fur-nissii***	***V. algino-lyticus***	***V. P. haemo-lyiticus***	***V. vul nificus***
Standard + 1% NaCl	73	1	24	0	33	0	0	8	0	0
1 % NaCI-Barritt method	93	9	96	0	95	0	0	83	0	0
Arginine dihydrolase Moeller's	0	0	0	0	81	52	0	0	0	0
Moeller's + 1% NaCl	0	**59**	0	95	93	100	0	0	0	0
Lysine decarboxylase Moeller's	99+	99	13	0	0	0	0	29	63	85
Moeller's + 1% NaCl	99+	100	36	0	52	0	0	99	100	99
Ornithine decarboxylase										

Table Contd

Table Contd.

Test	Positive for Vibrio species:									
	V. chol-erae	*V mime-cus*	*V. metschni-kovii*	*V. holli-sae*	*V. dam-sela*	*V. fluvi-alis*	*V. fur-nissii*	*V. algino-lyticus*	*V. P. haemo-lyiticus*	*V. vul nificus*
Moeller's	99+	99	0	0	0	0	0	3	71	47
Moeller's + 1% NaCl	99+	99	0	0	0	0	0	53	89	53
Gelatin hydrolysis										
Standard	52	60	65	0	5	41	56	45	66	55
Standard + 1% NaCl	62	63	38	0	6	85	86	76	89	75
Esculin hydrolysis										
Standard	0	0	4	0	0	0	0	0	0	0
Standard + 1% NaCl	2	0	59	0	0	8	0	2	1	39
Nitrate- *nitrite										
Standard	99+	99+	0	NG	NG	59	33	NG	0	2
Standard + 1% NaCI	99+	100	0	100	100	100	100	100	100	100

TABLE 16.6. KEY CHARACTERISTICS OF V. CHOLERAE AND V. MIMICUS

Properties	Reactions of	
	V. cholerae	V. mimicus
Shared by both species		
Oxidase	+	+
Growth in nutrient broth		
No NaCl added	+	+
1% NaCl added	+	+
Arginine dihydrolase		
Lysine decarboxylase	+	+
Ornithine decarboxylase	+	+
Fermentation of		
Lactose (1 to 2 days)	9	79
Lactose (3 to 7 days)	24	56
To differentiate the two species		
Sucrose fermentation	>99	0
Lipase (corn oil)	95	10
Voges-Proskauer	65	0

Changes were necessary in some of the original procedures, e.g., suspension of the organism in NaC1 solution (0.85 to 1%) rather than distilled water. The halophilic *Vibrio* species grew very poorly without the added NaCl and were often Ivsed in the distilled water. MacDonald et al. found that artificial seawater was better (gave more positive reactions) as a suspending medium for the halophilic *Vibrio* species. One problem has been that some of the commercial products confuse *Aeromonas* species with *V. fluvialis*. These two organisms can easily be differentiated with one additional test: *V fluvialis* is a halophilic *Vibrio* species and does not grow in nutrient broth without added NaCl, but *Aeromonas* strains grow in this medium. In summary, the kits have improved in their ability to identify *Vibrio* species, but misidentifications are still a problem. One of us (M.T.K.), who has tried several different kits, prefers conventional tube tests for *Vibrio* identification.

ANTIBIOTIC SUSCEPTIBILITY

Vibrio strains important in clinical microbiology usually grow well on Mueller-Hinton agar, which is used in the disk susceptibility test.

Although MuellerHinton agar contains no added NaCl (see the commercial formula), it contains a hydrochloric acid hydrolysate of casein, which apparently has enough NaCI to allow good growth of the halophilic *Vibrio* species. Some of the environmental mtfine vibrios grow poorly on Mueller-Hinton agar, presumably because of their higher requirement for Na^+. Broth dilution susceptibilities in Mueller-Hinton broth can be done without modification for most *Vibrio* species because the broth contains sufficient Na^+.

TABLE 16.7. DIFFERENTIATION OF THE CLASSICAL AND ELTOR BIOGROUPS OF V. CHOLERAE SEROGROUP 01

Property	*Biogroup*	
	Classical	*Eltor*
Voges-Proskauer	—	+
Zone around polymyxin B (50 U)	+	—
Hemolysis of erythrocytes	—	+
Agglutination of chicken erythrocytes	—	+
Lysis by bacteriophage		
Classical IV	+	—
Eltor 5	—	+

TABLE 16.8. DIFFERENTIATION OF THE ARGININE-POSITIVE SPECIES *V. DAMSELA, V. FLUVIALIS,* AND *V. FURNISSII*

	Reaction of Vibrio species		
Test	V. *damsela*	V. *fluvialis*	V. *furnissii*
Voges-Proskauer	+	—	—
Citrate, Simmons	—	+	+
L-Arabinose fermentation	—	+	+
D-Mannitol fermentation	—	+	+
Sucrose fermentation	—	+	+
Gas production during fermentation	V	—	+

Antibiotic resistance is rare in *Vibrio* species compared with species of the family *Enterobacteriaceae*. In most cases, resistance is probably intrinsic to the species rather than acquired through plasmid transfer or antibiotic exposure. The one exception to this generalization is the antibiotic resistance found in some outbreaks of *V. cholerae* which have become resistant through the acquisition of R factors. In the United States, strains of *V. cholerae* and other *Vibrio* species have rarely had this type of resistance. Resistance to polymyxin antibiotics (polymyxin B and colistin) can be useful in spotting a culture of the eltor biogroup of *V. cholerae* or a culture of *V. vulnificus*. Most other *Vibrio* species are more susceptible.

SEROLOGICAL TYPING OF ISOLATES

The serological typing of isolates is usually done by reference laboratories rather than by clinical laboratories. The one exception is the use of *V. cholerae* 01 antisera which can be used in clinical laboratories to test a strain of *V. cholerae* to determine whether the strain is 01 positive or negative. This procedure was discussed above. Strains of *V. cholerae* which are not 0 group I can be further differentiated by serotyping. Two systems for typing non-O1 *V. cholerae* have been described.

TABLE 16.9. DIFFERENTIATION OF *V FLUVIALIS* AND *V. FURNISSII*

Property	Positive for property	
	V.fluvialis	*V.furnissii*
Glucose, gas production	0	99
Esculin hydrolysis	72	0
Carbon sources, growth onCitrulline	97	4
Glucuronate (D-)	94	7
Putrescine	31	100
Aminovalerate (delta)	0	63
Cellobiose	63	4
Glutarate	-b	+b

In a system described by Sakazaki and Shimada, antisera are made against heated cells, and the 0 antigen is determined by agglutination. There are over 60 0 antigens in this schema. In a system described by Smith, antisera are made against unheated cells and the "Smith type" is determined by slide agglutination of living cultures. There are over 72 different types in the Smith schema. Serological typing of non-O1 *V. cholerae* should be done only to answer specific epidemiological questions, and such typing is done by only a few reference laboratories. An international working group will study the two systems and make recommendations for standardization.

There is also a serological typing schema for *V. parahaemolyticus*. This schema includes 11 numbered 0 antigens and 55 numbered K antigens. The 0 antigen is first determined by slide agglutination with an autoclaved antigen. The K antigen is then determined by slide agglutination with an unheated suspension.

DIAGNOSIS OF VIBRIO INFECTIONS

The best method for the diagnosis of infections due to *Vibrio* species is the isolation of the causative organism during the acute phase of illness. However, it is not always possible to obtain the necessary cultures, so an alternative is to measure the patient's antibody response to the *Vibrio* species which is suspected. This approach has been used frequently for cholera, and the methods have been discussed in detail by Feeley and DeWitt in the *Manual of Clinical Immunology*.

A retrospective diagnosis of cholera can be established with a high degree of certainty by the titration of acute- and convalescent-phase sera in agglutination, vibriocidal, or antitoxin tests. Acute-phase sera should be collected at 0 to 3 days, and convalescentphase sera should be collected at 10 to *21* days after onset of illness. Paired sera should be submitted to laboratories familiar with these specialized tests. A fourfold or greater rise in vibriocidal or agglutination titer is diagnostic, provided recent (within 2 weeks) cholera immunization can be ruled out. Antitoxin determinations may be of special value in persons recently immunized. The titration of single convalescent sera is usually not worthwhile, except that the absence of significant levels of vibriocidal antibody would nearly exclude the diagnosis of cholera.

Similar serological tests can be used with the other *Vibrio* species and are especially recommended to further document the etiological role of some of the new species. Antibody responses against the patient's particular strain can be particularly useful in distinguishing colonization from actual infection and can be determined in both intestinal and extraintestinal infections.

V. CHOLERAE SEROGROUPS 01 AND NON 01

In the mid-1930s it was recognized that the vast majority of vibrios isolated from cholera cases agglutinated in a single antiserum.

Agglutination in this antiserum (later called *V. cholerae* 01 antiserum) became the main criterion for identifying an organism as *V. cholerae*. Organisms which did not agglutinate in this serum were given the vernacular names "nonagglutinating vibrios," "NAGs," or "noncholera vibrios." These names included organisms known today as *V. cholerae* non 01, *V. parahaemolyticus,* other *Vibrio* species, and even *Aeromonas* or *Plesiomonas* species. It was soon recognized that some of the NAGs were identical with *V. cholerae* except for their agglutination in the antiserum to serogroup 01. However, it has been only in the last few years that these vibrios have been classified in the species *V. cholerae*. Today it is accepted that there are two groups of *V. cholerae* strains, i.e., serogroups 01 and non 01. It is usually convenient to discuss these serogroups separately. *V. cholerae 01 is* the serogroup isolated from cholera cases, in which it produces severe, watery diarrhea through the action of cholera toxin. *V. cholerae* non 01 can cause a choleralike illness, but also causes a much wider spectrum of disease, including extraintestinal infections.

V. cholerae serogroup 01 is the group of *V. cholerae* strains which cause pandemic cholera. These strains are defined as strains of *V. cholerae* which agglutinate in 01 serum, include serogroups Ogawa and Inaba (and Hikojima), and include biogroups classical and eltor.

History

Historians do not always agree on the history of cholera, perhaps because early descriptions made it difficult to differentiate cholera from other secretory diarrheas. The cholera bacillus was isolated by Koch in 1884, so bacteriological confirmation was possible only after this date. Cholera, or Asiatic cholera as it was often called in early writings, has probably always existed in India. It may have been found outside of India before 1800, but not in large epidemics. The spread of Asiatic cholera has come in seven large pandemics. The first pandemic began in 1816 to 1817; six other pandemics followed, beginning in 1829, 1852, 1863, 1881, 1889, and 1961. Until recently, the classical biogroup has caused pandemic cholera. However, the seventh pandemic was caused by the eltor biogroup. Recently the classical biogroup has displaced the eltor biogroup in some areas of India.

Clinical significance

In an area in which cholera is endemic, many individuals will ingest *V. cholerae*. Most of these people will have either a mild diarrhea or no symptoms at all (asymptomatic colonization of the intestine with recovery of the organism from formed stool). This pattern has been much more common in the seventh cholera pandemic caused by the eltor biogroup of *V. cholerae*. At the other extreme, some individuals develop a dramatic, acute diarrhea which has been called cholera gravis. This is the terrifying disease which caused such fear in the previous century when the word cholera was mentioned.

In severe cholera there is massive diarrhea, with large volumes of rice-water stool (clear fluid with flecks of mucus) passed painlessly. The amount of fluid passed can be 1 liter per hour. In 4 to 6 days this fluid would amount to over twice the body weight. There is usually vomiting and little desire to maintain nutrition through normal eating. If the patient is untreated, there will be prostration with symptoms of severe dehydration: painful muscle cramps, watery eyes, loss of skin elasticity, and anuria (absence of urine excretion). Death can occur very quickly after onset of symptoms because of the severe dehydration.

The mainstay of treatment is intravenous therapy, which restores water and electrolyte balance and prevents acidosis. Cases with milder illness are often treated with oral electrolyte solutions. Tetracycline therapy reduces the period of excretion but is not a substitute for rehydration.

Milder forms of diarrhea due to *V. cholerae 01* are more difficult to distinguish from other mild diarrheas. These cases can last from I to 5 days with several liquid stools per day. Cramping and vomiting can also occur. The role of *V. cholerae 01 is* shown by the isolation and identification of the causative organism. With the eltor biogroup of *V. cholerae 01,* this milder disease is seven times more common than severe cholera.

Pathogenesis

The pathogenesis of diarrheaa due to *V. cholerae 01 is* well understood. The organism is ingested, and some cells survive the acid pH of the stomach and pass into the small intestine. The organism colonizes the small intestine and begins to grow and produce cholera toxin. Cholera toxin (molecular weight, 84,000) is composed of an Al subunit (molecular weight, 21,000), an A2 subunit (molecular weight, *7,000),* and five B subunits (molecular weight, 10,000 each). The B subunit of intact cholera toxin attaches to a specific receptor, ganglioside G1, on the cell membrane of the host's tissue. The Al subunit of cholera toxin then activates the enzyme adenylate cyclase in the host cells, which then increases the level of cyclic AMP and the hypersecretion of salt and water. The net results are the massive outpouring of liquid stool and resulting dehydration which are typical of cholera.

Sources

V. cholerae serogroup 01 is well adapted to the human intestinal tract. The vast majority of strains come from human feces Only a few human isolates are extraintestinal. *V. cholerae* serogroup 01 can also be isolated from environments and from other animals that come in contact with feces of cholera cases. There have been reports of *V. cholerae* serogroup 01 from environmental sources which are unlikely to have come from cholera cases. These strains do not produce cholera toxin and probably represent different clones of *V cholerae.* They are probably similar to the *V. cholerae* non-O1 strains in clinical and public health importance.

Isolation

In rice-water stools from cholera cases, *V. cholerae is* usually present essentially as a pure culture and in very high numbers (10^6 to 10^8 organisms per ml of stool). Isolation will present no problem in these cases. In formed feces, the organism will probably be present in much lower numbers, and normal enteric flora will also be present. The approach will yield more positives in this situation.

Biochemical reactions

In countries in which cholera is common, there is no need to do extensive biochemical studies to confirm a culture as *V. cholerae.* However, complete biochemical testing is recommended in industrialized countries where cholera is rare. The test for Na+ requirement differentiates *V. cholerae* from the halophilic *Vibrio* species. Sucrose fermentation differentiates *V. cholerae* and *V. mimicus*. The decarboxylase pattern of arginine negative, lysine positive, and ornithine positive differentiates *V. cholerae* from the oxidase-positive fermentation species in the genera *Aeromonas* and *Plesiomonas*.

Antibiotic susceptibility

Antibiotic resistance is rare in *V. cholerae*. Most strains are susceptible to tetracycline, the drug of choice. Occasionally a strain of *V. cholerae* becomes resistant and can spread. This event usually happens where cholera is endemic and there is more chance for a strain to acquire an antibiotic resistance plasmid.

Susceptibility to the polymyxin group of antibiotics is a property used to differentiate the classical and eltor biogroups.

Geographical distribution

The World Health Organization regularly reports on the worldwide distribution of cholera. In its survey for 1982, 54, 856 cases were reported from *37* countries. The majority of cases were in Africa, Asia, and the Trust Territories in the Pacific Ocean.

Cholera (defined as the disease caused by *V. cholerae* serogroup 01, cholera toxin positive) in the United States is quite interesting. In the United States, the last case of cholera from a pandemic occurred before 1900. Then, unexpectedly, an isolated case occurred in Texas in 1973. This case was followed by an outbreak in Louisiana in 1978, two sporadic cases in Texas in 1982, and an outbreak on an oildrilling platform on a bayou near the Texas coast in 1982. These strains have been thoroughly studied and have the following properties in common: serogroup Inaba, cholera toxin positive, strongly hemolytic, and a unique bacteriophage lysis pattern. When compared with other cultures from the seventh pandemic of cholera, the American cultures are quite different, suggesting that they are a different strain (clone). The origin and history of this strain are now being studied.

***V cholerae* Serogroup Non 01**

V. cholerae serogroup non 01 can cause a choleralike disease, but it is also isolated from patients with mild diarrhea and extraintestinal infection. It is also isolated from the aquatic environment.

History

Until recently, the strains of serogroup non 01 were not classified in the species *V. cholerae* but were reported as "nonagglutinating vibrios," "NAGs," or "noncholera vibrios." Today they are classified as *V. cholerae* non 01, and the literature on their ecologyand role in human disease is expanding rapidly.

Clinical significance: diarrhea

These organisms have been isolated from patients with severe dehydrating (choleralike) gastroenteritis. Other strains have been isolated from patients with mild diarrhea. Unlike *V. cholerae 01,* the non-O1 strains have also been found in patients with diarrhea when feces have contained blood or mucus or both. Fever is also a symptom in some of the cases. These different responses may correlate with the different types of pathogenesis described below. The duration of illness is usually less than *3* days.

Pathogenesis of intestinal infections

Although the pathogenesis of intestinal infections is not as well known as is that of *V. cholerae 01,* three pathogenic mechanisms have been postulated: (i) production of cholera (choleralike?) toxin, (ii) production of a heat-stable enterotoxin (positive result in the infant mouse assay used to detect *Escherichia coli* heatstable enterotoxin), and (iii) invasive disease (positive response to whole cells but not to cell filtrates in the ligated rabbit ileal loop, or infant mouse assay). Some strains of *V. cholerae* non 01 have no positive response in any of the above assays. The correlation of the possible virulence factors with the different types of gastroenteritis-invasive diarrhea needs further investigation.

Extraintestinal infections

V. cholerae non 01 has a wider spectrum of disease than does *V. cholerae 01*. Non-O I strains have been isolated from patients with septicemia who had cirrhosis or other underlying disease. In our series, there were 42 blood isolates, compared with only 1 blood isolate for *V. cholerae 01*. The non-O1 strains were also isolated from ears, wounds, respiratory tract, and urine. There were 52 isolates of serogroup non 01 from these sources, compared with only 2 isolates for *V. cholerae 01*.

Laboratory

The discussion of laboratory procedures for *V. cholerae 01* (see above) also applies to the non-O1 strains. The essential difference is that the 01 strains agglutinate in *V. cholerae* serogroup 01 antiserum, but the non-01 strains do not. This difference emphasizes the need for reference laboratories to maintain this antiserum, even though it will be needed only occasionally. The biochemical reactions of both *V. cholerae* groups are almost identical.

Geographical distribution

V. cholerae non-O1 strains have been isolated worldwide both from patients and from the environment. In the United States, many strains are found in sporadic cases of diarrhea. Most of these cases have occurred after raw oysters were eaten. In the United States, fewer than 5% of the isolates have been positive for cholera toxin, a situation in contrast to that in Bangladesh, where about one-third of the isolates have been positive.

VIBRIO MIMICUS

V. mimicus is a new *Vibrio* species which is associated with diarrhea, usually after the consumption of uncooked seafood, particularly raw oysters *V mimicus* has been isolated in many countries, which suggests a worldwide distribution in countries situated on an ocean. The isolation of *V. mimicus* from water, oysters, and shrimp and its distribution in coastal areas indicate that its ecology may be similar to the ecology of *V. cholerae* non 01, which has been well studied.

History

There have been a number of reports on *V cholerae* strains with atypical biochemical reactions. Traditionally, these strains have been reported with designations such as "*V. cholerae* lysine decarboxylase negative," "*V. cholerae* mannitol negative," or "*V. cholerae* sucrose negative." Davis et al. studied representative strains from six of these unusual biogroups by both DNA hybridization and phenotypic analysis. Five of the six groups were very highly related to *V. cholerae* by DNA hybridization. These groups had been identified correctly as atypical strains of *V. cholerae*. However, the group of sucrosenegative strains was only 24 to 54% related to *V. cholerae*. Based on this low relatedness and phenotypic differences, Davis et al. proposed a new species, *V. mimicus*. The species name (specific epithet) *mimicus* refers to the fact that the strains "mimic" *V. cholerae*.

Clinical significance

Based on the information submitted with cultures, Davis et al. felt that *V. mimicus* was associated with diarrhea and was probably linked to the eating of shellfish. Three strains of *V. mimicus* were positive for heat-stable enterotoxin in the infant mouse assay, and five strains were positive for heat-labile enterotoxin in the Y-1 adrenal cell assay or the enzyme-linked immunosorbent assay. However, other strains from diarrhea cases were negative in the assays. The enterotoxin data were very similar to those for non-O1 *V. cholerae* strains in that some produced heat-labile or heat-stable enterotoxin, but most strains from patients with diarrhea produce neither toxin.

Shandera et al. did a retrospective case-control study on *17* patients whose cultures of *V. .mimicus* had been sent to CDC from *1979* to *1981*. This study further strengthened the association of *V. mimicus* with gastroenteritis-diarrhea after the consumption of seafood, particularly raw oysters. The role of *V. mimicus* as an actual cause of diarrhea needs stronger documentation.

Ear infection

Two of the *V. mimicus* isolates studied by Shandera et al. were from patients' ears. Case 1 was a 39year-old person with chronic otitis media since the age of 20. Case 2 was a 13-year-old child who had Kawasaki's syndrome at age 5. Both patients had seawater exposure during the week before onset of illness, one person wading in a North Carolina estuary and the other swimming off the coast of Hawaii. These data for *V. mimicus* are similar to those for other *Vibrio* species from ear infections. *V mimicus* has been recovered from ears after salt- or brackish-water exposure, but its causal role in ear infections needs further documentation.

Isolation

Most of the *V. mimicus* strains were isolated during a search for *V. cholerae* and other *Vibrio* species. *V. mimicus* is sucrose negative, so colonies on TCBS agar are green rather than yellow. It grows well on normal enteric media, so isolation and identification should pose no special problems.

Biochemical reactions

These two species *V. mimicus* and *V. cholerae* are very similar in most of the tests normally needed for identification; however, given Table gives reactions which are useful for differentiation. *V. mimicus is* negative for lipase production and sucrose fermentation. *V. cholerae* usually has the opposite pattern, although a few strains are lipase negative. *V. mimicus* resembles the classical biogroup of *V cholerae* because it is Voges-Proskauer negative and susceptible to polymyxin. The eltor biogroup of *V. cholerae is* usually Voges-Proskauer positive and resistant to polymyxin. *V. mimicus* grows in nutrient broth without NaCI, a reaction which differentiates it along with *V. cholerae* from the halophilic *Vibrio* species which require a higher NaCl concentration for growth.

Antibiotic susceptibility

V. mimicus is generally susceptible to antibiotics except for sulfadiazine *(87%* resistant). A few strains are resistant to penicillins, colistin, nalidixic acid, and kanamycin. V. mimicus. is v*ery* similar to *V. cholerae* in its susceptibility.

VIBRIO PARAHAEMOLYTICUS

V. parahaemolyticus has been known as a cause of acute gastroenteritis since 1950. Food-borne outbreaks and sporadic cases occur worldwide and are usually associated with the consumption of contaminated seafood. Two books and a recent review article summarize much of the information known about this organism.

History

On 20 and 21 October 1950, there was an outbreak of food

poisoning in Osaka, Japan. There were 272 patients with acute gastroenteritis, 20 of whom died. These deaths led to an extensive investigation of the outbreak, and eventually an organism was isolated which was shown to be the etiological agent. The properties of the organism were studied, and it was named as a new bacterial species, *Pasteurella parahaemolytica*. The halophilic nature of the new organism was not discovered until 1955, and it was then reclassified as a halophilic *Vibrio* species, *V. parahaemolyticus*.

Clinical significance in gastroenteritis

V. parahaemolyticus causes gastroenteritis with nausea, vomiting, abdominal cramps, low-grade fever, and chills. The diarrhea is usually watery but sometimes bloody. The disease is usually mild and self-limiting but can be fatal (a 7% fatality rate in the first reported outbreak). There is a good correlation between pathogenicity and a positive Kanagawa test which measures the ability of the strain to produce a hemolysin for human erythrocytes when it is grown on a special medium. About 96% of the *V. parahaemolyticus* strains from well-documented cases of human gastroenteritis are Kanagawa positive, but only about 1 % of the strains isolated from the environment are positive. Although there is excellent correlation between this hemolysin and human disease, the mechanisms of pathogenesis are still being debated, since other toxins or virulence factors may be involved. Rehydration is usually the only treatment needed, but in some severe cases, the patient will require hospital admission. Antimicrobial therapy may be beneficial, and tetracycline appears to be the drug of choice.

Outbreaks of gastroenteritis due to *V. parahaemolyticus* occur worldwide, but they are not as common in the United States as in some other countries. In Japan, *V. parahaemolyticus* is an extremely important diarrheal agent and causes 50 to 70% of the cases of foodborne enteritis. All of the outbreaks are associated directly or indirectly with seafood. Direct infection comes from the ingestion of raw fish or shellfish which are contaminated with *V. parahaemolyticus*. This mechanism of transmission is apparently quite common in Japan because of the national custom of eating raw fish. Contamination after cooking is apparently the mechanism for indirect infection. This mechanism is important in outbreaks where the food has been cooked. Kitchen utensils and water probably introduce the organism into the cooked food, where it can then multiply if correct food-handling procedures are not used.

Extraintestinal infections

There is confusion about the role of *V. parahaemolyticus* in infection outside the intestinal tract. Many authors refer to rapidly fatal septicemia or to wound infections due to this organism. However, in many of the original reports the organism was actually *V. vulnificus* rather than *V. parahaemolyticus*. This error is due to the fact that *V. vulnificus* was incorrectly identified as *V. parahaemolyticus*. All reports of severe extraintestinal infections due to *V. parahaemolyticus* must be examined critically to be sure the organism in question is not *V. vulnificus,* another *Vibrio* species, 'or a marine vibrio. Two cases of eye infections provide evidence that *V. parahaemolyticus* can be a pathogen outside the intestinal tract. However, these cases are apparently rare. The vast majority of the isolates in our series are from feces, but 2 were from blood and 11 were from wounds. Isolates from the marine environment are quite common), a fact which conforms to the known distribution of the organism.

Isolation

Unlike the situation with most of the *Vibrio* species, much has been written about the isolation of *V. parahaemolyticus*. The fact that this species was not detected as an important cause of gastroenteritis until 1950 indicates that isolation with the usual enteric media can pose problems. Oxidase testing of colonies on a nonselective medium such as blood agar should provide a high isolation rate. A plate of TCBS medium can be included in those geographical areas where *V. parahaemolyticus* is most common. The isolation methods designed for *V. cholerae* are also efficient for *V. parahaemolyticus*. *V. parahaemolyticus* grows well on TCBS agar as 2- to 3-mm green colonies. A number of selective media have also been described elsewhere, but these media are most useful with environmental specimens or for detecting carriers and are almost never used in. the routine processing of clinical specimens.

Identification

V. parahaemolyticus is one of the halophilic *Vibrio* species and belongs in the lysine-positive, argininenegative group. It is negative for the VogesProskauer test and for the fermentation of lactose and salicin. These and other tests are useful in differentiation.

Urea-positive strains of *V. parahaemolyticus*

Until recently, *V. parahaemolyticus* was considered negative for urea hydrolysis. Sakazaki and Balows reported that none of their 2,354 strains was urea positive. However, in the last few years we have received strains which were otherwise typical of *V. parahaemolyticus* but were urea positive. These ureapositive strains were confirmed by DNA hybridization as *V. parahaemolyticus*. The number of urea-positive strains has increased steadily: in 1977, 17% were urea positive; in 1978, 14%; in 1979, 36%; in 1980, 12%; in 1981, 47%; in 1982, 73%; and in 1983, 83%. This sample is biased because laboratories refer atypical strains rather than typical ones. However, 50% or more of *V. parahaemolyticus* strains from California and from outbreaks investigated by CDC have also been urea positive. These samples are less biased and show that the increase in urea-positive strains is real.

Antibiotic susceptibility

V. parahaemolyticus is usually resistant to ampicillin and car

benicillin but susceptible to colistin. These properties are shared with *V. alginolyticus,* but V. *vulnificus* has the opposite susceptibilities).

VIBRIO HOLLISAE

V. hollisae is a recently described halophilic species in the genus *Vibrio* which is associated with diarrhea after the consumption of raw seafood.

History

V. hollisae was named as a new species in 1982. Two laboratories had independently studied this organism under the vernacular names "EF 13" and "Enteric Group 42." Data from DNA hybridization indicated that all these strains were highly related and distinct from other named *Vibrio* species. Hickman et al. named this organism *V. hollisae.*

Clinical significance: sources

V. hollisae is a probable cause of human diarrhea. Its association with diarrhea is very strong, but additional evidence is needed to document its causal role and pathogenesis. Of the original 16 strains, 15 were from feces, and many of the patients had diarrhea. Morris et al. described the clinical and epidemiological features of 11 diarrhea cases that were culture positive for *V. hollisae*.

They also reported one isolate from blood. It was from a patient with hepatic cirrhosis, hepatic encephalopathy, bronchopneumonia, and sepsis due to the yeast *Cryptococcus*. The patient was comatose and died 2 days after hospital admission.

Isolation

Laboratory strains of *V. hollisae* did not grow (2 days, 36°C) on TCBSor MacConkey agar. Since these two media are frequently used for the isolation of *Vibrio* species and species of the family *Enterobacteriaceae* from stools, it is possible that *V. hollisae is* often missed, even though it may be the predominant organism. *V. hollisae* grows well on sheep blood agar (36°C), so it can be detected by the use of the oxidase reagent as discussed above.

Identification

V. hollisae does not grow in nutrient broth without added NaCl; thus it is one of the halophilic *Vibrio* species. Only a few reactions need be mentioned. *V. hollisae* is indole positive when tested in heart infusion broth with 1% NaCl added, but only 38% positive in peptone water with only 0.5% NaCl. Moeller's lysine, arginine, and ornithine are all negative. Motility in *V. hollisae* is very unusual. None of the strains was motile (semisolid medium, 36°C) after 48 h of incubation, but after 7 days, 88% were motile. This characteristic is unusual for the genus *Vibrio*. *V. hollisae* ferments only D-glucose, L-arabinose, D-galactose, and D-mannose, a characteristic fermentation pattern among *Vibrio* species. Recently we confirmed a ureapositive strain of *V. hollisae,* which is the first one with this unusual property.

Antibiotic susceptibility

Stains of *V. hollisae* have a very characteristic antibiogram. There are very large zones around all antibiotics tested, including penicillin.

VIBRIO FL UVIALIS

V. fluvialis is a new *Vibrio* species which seems to cause sporadic cases of diarrhea worldwide and has been implicated in outbreaks of diarrhea in Bangladesh.

History

In 1981, Lee et al. gave the name *V. fluvialis* to a group of halophilic vibrios which had been previously known as "group F vibrios" and as "group EF6." These organisms had been isolated from a number of environmental sources throughout the world. They had also been isolated from humans with diarrhea. Lee et al. defined two biogroups (synonym, biovars) in this species which correlated with source and certain biochemical tests. *V. fluvialis* biogroup was anaerogenic and was isolated from human diarrhea as well as from the environment. *V. fluvialis* biogroup II produced gas during fermentation (it was aerogenic) and was isolated from the environment, but not from humans with diarrhea. Brenner et al. later showed that strains of *V. fluvialis* biogroup II were more distantly related to strains of *V. fluvialis* biogroup I and proposed that these strains be classified in a new *Vibrio* species, *V. furnissii.*

Clinical significance: sources

The sources of the early clinical isolates of *V. fluvialis* showed a marked association with diarrhea. Further studies have provided additional evidence for its causal role. One difficulty in assessing the clinical significance and epidemiology of *V. fluvialis* has been the presence of other possible pathogens. Most of the reports have come from geographical areas where several possible pathogens are often present in feces.

Gastroenteritis caused by *V. fluvialis* is usually described as choleralike. Patients typically have watery diarrhea with vomiting (97%), abdominal pain (75%), moderate to severe dehydration, and often feve. Usually infants, children, and young adults are affected. Frank blood is found in a small percentage of stool samples, but erythrocytes or leukocytes are found in most.

The pathogenesis of *V. fluvialis* diarrhea is not completely known. Assays for heat-labile and heatstable enterotoxins and invasiveness have generally been negative. About 20% give a positive rabbit ileal loop test for enterotoxin. Recent studies have provided additional evidence for enterotoxin or enterotoxinlike molecules.

Isolation

V. fluvialis grows on TCBS agar as 2- to 3-mm yellow colonies. Although this species requires Na+ its requirement is much less than some of the more halophilic *Vibrio* species, which makes isolation and identification easier. Nishibuchi and Seidler recently described a selective broth for isolating this species from water.

Identification

The most striking aspect of *V. fluvialis* identification is the possible confusion with *Aeromonas* strains, since both species are arginine dihydrolase positive. Many cultures we receive as "*V. fluvialis* " turn out to be *Aeromonas hydrophila*. The converse is also true. A simple .test to differentiate these two organisms is growth in nutrient broth with 0 and 1% NaCl. *V. fluvialis is* a halophilic vibrio and will not grow without the added NaCl. A. *hydrophila* grows in both media.

VIBRIO FURNISSII

V. furnissii is a new *Vibrio* species which was formerly known as "*V. fluvialis* biovar II," "*V. fluvialis* aerogenic," or "*V. fluvialis* gas'." It is now recognized as a separate species and is named *V. furnissii).*

Clinical significance: sources

The vast majority of the isolates of the *V. fluvialis- V. furnissii* group isolated from patients with diarrhea have not produced gas during fermentation. Thus, they are *V. fluvialis*. One exception was a case of human diarrhea in Indonesia. Further investigation into case histories of the isolates studied at CDC indicated that *V. furnissii* had been isolated from two outbreaks of acute gastroenteritis. These outbreaks were among American tourists returning from the Orient in 1969. Other enteric pathogens were also isolated *(V parahaemolyticus, V. cholerae* non 01, *V. fluvialis, Salmonella spp.*, and *Plesiomonas spp.), so* the causal role of *V. furnissii* was not clear. There have been no documented cases of diarrhea. caused by this organism. Thus we have not included it with the other *Vibrio* species as a probable cause of diarrhea. However, microbiologists should be alert to its presence and for evidence of its causal role in diarrhea. *V. furnissii* is apparently rare in human clinical specimens. Lee et al. mention that *V. furnissii* (listed as *V. fluvialis* biovar II) is widespread in the aquatic environment and is common in estuaries.

Laboratory

V. furnissii resembles *V. fluvialis* very closely. The two species are so close phenotypically that only a few tests are available to differentiate them. Gas production will be the key differential test in most clinical laboratories. *V. furnissii* can be confused with strains of *Aeromonas* if its halophilic nature is not recognized.

VIBRIO VULNIFICUS

V. vulnificus has been recognized as a distinct species of *Vibrio* since 1976. It causes wound infections and life-threatening septicemia. In the last few years, there has been intensive study of this organism.

History

In 1976, Hollis et al. described a salt-requiring organism that appeared to be different from other *Vibrio* species. Its properties were similar to those of *V. parahaemolyticus* and *V. alginolyticus,* but the new organism fermented lactose. The vernacular name "lactose-positive *Vibrio* " was given to this organism. In the literature it has been referred to as "L$^+$ *Vibrio* " and "Lack *Vibrio.* " This organism was studied by several laboratories and given the scientific name *Beneckea vulnifica* by Reichelt et al. in 1976. Its classification in the genus

Beneckea was not widely accepted, so it was subsequently classified in the genus *Vibrio*. Today, it is almost universally known as *V. vulnificus*.

Clinical significance: sources

V. vulnificus has been associated with two disease syndromes, primary septicemia and wound infection.

Primary septicemia is a very serious infection with a fatality rate of about 50%. Most patients with primary septicemia due to *V. vulnificus* have preexisting liver disease, but other patients have been normal individuals. In most cases the disease begins several days after the patient has eaten raw oysters. Individuals develop malaise which progresses rapidly to fever, chills, and prostration and often to severe hypotension and death. The progression can be very rapid, from asymptomatic to death within 24 h. At autopsy, there are findings of gram-negative shock, hemorrhage, and necrosis of involved muscles and skin (with organisms e culturable from the lesions). Cultures of blood and skin lesions are usually positive for *V. vulnificus*.

V. vulnificus also causes severe wound infections, usually after trauma and exposure to marine animals or the marine environment . The mortality rate is not nearly as high as in primary septicemia caused by this. organism (about 7% compared with about 50%, respectively). Wound infections progress rapidly with swelling and erythema, often followed by the development of bullae or vesicles leading to tissue necrosis. Symptoms include a very painful wound, fever, and chills. Surgical intervention, including amputation, is often required. Antimicrobial therapy is often required; the most effective drugs appear to be tetracycline and aminoglycosides, based on laboratory susceptibilities. About onethird of the patients with wound infection had underlying disease.

Isolation

V. vulnificus grows well on blood agar and TCBS agar. Although most strains of *V. vulnificus* are sucrose negative (3% positive in the original report of Hollis et al., the number of sucrosepositive strains in our collection has increased markedly over the years. Thus, on TCBS agar, *V. vulnificus* can be either green (most) or yellow (some). This fact has not always been considered in searches for this organism. Most isolations of *V. vulnificus* in clinical laboratories will be made on blood agar, since blood and wounds are the usual sources of this species. Commercial blood culture bottles usually support growth of *V. vulnificus;* however, the Na^+ content is probably suboptimal, which may yield some cells with aberrant sizes and shapes. *V. vulnificus* does not survive well on media with an NaCl content of only 0.5%. Some cultures are dead by the time they reach our laboratory.

Identification

V. vulnificus is a halophilic *Vibrio* species. It is lysine positive and arginine negative, which puts it into the same' group as *V. parahaemolyticus* and *V. alginolyticus*. It is differentiated by lactose fermentation, the ONPG test, salicin fermentation, and several other tests. The antibiogram (disk diffusion) is also helpful in the differentiation of these three species. *V. vulnificus* has either no zone or a small zone around the colistin disk, but large zones around ampicillin and carbenicillin. *V. parahaemolyticus* and *V. alginolyticus* have the opposite pattern.

Antibiotic susceptibility

V. vulnificus is susceptible to most antibiotics except the polymyxin group. Hollis et al. tested 33 strains by both the disk and agar dilution methods (MICs on Mueller-Hinton medium with no added salt). All strains were susceptible to penicillin, ampicillin, carbenicillin, cephalothin, chloramphenicol, gentamicin, tetracycline, rifampin, nitrofurantoin, and sulfisoxazole. Although *V. vulnificus* is susceptible to most antibiotics used against gramnegative bacteria, animal studies indicate that tetracycline may be most effective in vivo. Early administration of aminoglycoside or tetracycline antibiotics before the onset of hypotension appears to be critical to successful therapy.

Other infections

Other infections from which *V. vulnificus* has been isolated include pneumonia in a drowning victim and endometritis which developed in a woman after exposure to seawater.

Sources

The vast majority were from blood or wounds. Unusual isolates included one from urine and five from stool.

V. vulnificus is also isolated from the marine environment.

Pathogenesis

Animal studies have shown that *V. vulnificus* can cause severe local infections with gross edema leading to tissue necrosis and death. Iron availability has been implicated in the pathogenesis of *V. vulnificus* infections. A toxin produced by *V. vulnificus* has been demonstrated, although its role in the pathogenesis of infections is still unclear.

VIBRIO DAMSELA

V. damsela is a new *Vibrio* species which is associated with human wound infections. It is found in the marine environment and causes skin lesions on certain marine fish.

History

V. damsela was described in 1981 by Love et al.. The original isolates were from skin lesions on marine fish from off the California coast. Koch's postulates were fulfilled when it was documented that this organism caused the lesions on the damselfish.

Strains of *V. damsela* were compared with unidentified human strains of *Vibrio,* and several *V. damsela* strains were identified. The strains from marine fish, human clinical specimens, and the environment led to the proposal of this new species in the genus *Vibrio*.

Clinical significance: sources

Morris et al. reviewed the case histories of six patients with wound infections from which isolates had been sent to CDC from *1971* to 1981. All of the infections had been acquired in coastal areas.

All of the clinical isolates were from wounds; 9 of the 10 wounds were leg or foot wounds. Marine animals, particularly marine fish, made up the majority of the remaining isolates. A number of fish from Dakar, Senegal, yielded this organism. Other isolates included sewage (one isolate), oysters (one isolate), and a wound on a raccoon (one isolate). The role of *V. damsela* as an actual cause of human wound infection needs further documentation.

Isolation

V. damsela grows on blood agar and other noninhibitory media, but the Na+ content is less than optimum. Strains grow on TCBS agar as 2- to 3-mm green colonies. Love et al. noted that the damselfish isolate grew much better on TCBS at 25 than at 36°C. There have been few details on the isolation of this organism from human clinical specimens.

Identification

V. damsela is a typical halophilic *Vibrio* species in many respects. Strains do not grow in nutrient broth without added NaCl, which indicates its requirement for Na^+. *V. damsela is* arginine positive and Voges-Proskauer positive and has a characteristic fermentation pattern. It ferments D-glucose, D-mannitol, and D-mannose, and most strains ferment D-galactose and trehalose. The characteristics which differentiate it from other *Vibrio* species

Antibiotic susceptibility

V. damsela is generally susceptible to antibiotics, but less so than many other *Vibrio* species. Resistance is often seen for ampicillin, carbenicillin, penicillin, streptomycin, and sulfadiazine .

VIBRIO ALGINOLYTICUS

V. alginolyticus is a relatively new species in the genus *Vibrio*. Most clinical isolates have come from infected wounds, ears, or eyes after trauma and seawater exposure. It is very common in the marine environment.

History

V. parahaemolyticus was the first of the halophilic *Vibrio* species to be studied in detail. An organism similar to, but different from, *V. parahaemolyticus* was often isolated from the marine environment, but not from patients with gastroenteritis. This organism was originally called "biotype 2 of *V. parahaemolyticus*. " However, a number of studies have shown it to be a distinct species, i.e., *V. alginolyticus*.

Clinical significance: sources

There is now a considerable amount of literature reporting the isolation of *V. alginolyticus* from softtissue infections. Wound and ear infections are usually mentioned, with eye infections mentioned much less frequently. The data show the definite association of *V. alginolyticus* with infection at these sites; however, the etiological role and pathogenesis of this organism have not been thoroughly shown. However, most authors list *V. alginolyticus* as a pathogenic *Vibrio* species, particularly of wound and ear infections. Antibiotic treatment has been used in most cases, and surgical debridement has been used in some.

Wounds and ears are the most common sources. Interestingly, there were 12 isolates from foot or leg wounds. *V. alginolyticus* is-also widely distributed in the marine environment. Most clinical isolates probably come from patients exposed to the marine environment, particularly after trauma.

Isolation

V. alginolyticus is a marine *Vibrio* species and requires added NaCl for optimum growth. However, it grows well on many of the enteric media. It is sucrose positive and occurs as 2- to 3-mm yellow colonies on TCBS agar.

Identification

It is lysine positive and arginine negative, which puts it into the group with *V. parahaemolyticus* and *V. vulnificus*. *V. alginolyticus* is usually Voges-Proskauer positive and usually grows in 8 and 10% NaCl, which differentiates it from the other two species. Table 13 gives some other differential characteristics.

Antibiotic susceptibility

V. alginolyticus is usually susceptible to ampicillin and carbenicillin but resistant to colistin. This pattern is similar to that of *V. parahaemolyticus; V. vulnificus* has the opposite pattern.

VIBRIO METSCHNIKOVII

V. metschnikovii is an old species in the genus *Vibrio* that has been frequently isolated from fresh, brackish, and marine waters. It has been isolated from clinical specimens only rarely.

History

In 1884, Gamalria reported the isolation of a new organism, *V. metschnikovii,* from a fowl that had died of a choleralike disease. Little was written about the organism until Lee et al. began isolating similar organisms on TCBS agar from marine and freshwater environments in the United Kingdom. These strains were oxidase negative and did not reduce nitrate to nitrite, but otherwise they were quite typical halophilic *Vibrio* species. Lee ét al. proposed that the strains be classified as *V. metschnikovii* and wrote an emended description of this organism.

Clinical significance: sources

Lee et al. described 40 strains of *V. metschnikovii,* but none were from human clinical specimens; however, in 1981 JeanJacques et al. reported the first clinically significant isolate of this newly redefined species. An 82-year-old woman with peritonitis and an inflamed gallbladder yielded *V. metschnikovii* from a positive blood culture. The isolate was considered clinically significant.

The sources of the CDC isolates. One unusual finding was the two strains from urine. One was from the midstream urine of an 80-year-old woman in Canada. The other was from a routine urine culture of a 65-year-old male with chronic alcoholism, poorly controlled diabetes, and incontinence and urinary frequency. No information was available on the clinical significance of these two isolates. The first wound culture was from a 64-year-old female who was on chronic renal dialysis and had peripheral vascular disease. She developed an ulcer around a bunion on her foot which was treated by a wedge resection, followed by a foot amputation and finally a below-the-knee amputation. The wound yielded *V. metschnikovii* along with five other possible pathogens. The patient went fishing frequently, which may have been her source of *V. metschnikovii*. Further study is needed to determine the clinical significance of *V. metschnikovii* from sources such as urine and wound.

Nonhuman sources (taken from Lee et al. and our own data) include rivers (11 isolates), sewage (9 isolates), cockles (4 isolates), shrimp (3 isolates), lobster (2 isolates), crab (1 isolate), and fowl (2 isolates). These data suggest that *V. metschnikovii is* widely distributed in the environment and that humans occasionally become colonized or infected from these sources.

Laboratory

V. metschnikovii grows on TCBS agar as 2- to 3-mm yellow colonies and apparently grows well on other laboratory media. It requires Na+ for growth but in much smaller amounts (5 to 15 mM) than the other Na^+-requiring *Vibrio* species. Table gives the complete characteristics of *V. metschnikovii*. This species is very easy to identify because it is oxidase negative, does not reduce nitrate to nitrite, and requires Na+ for growth. These three properties differentiate it from all the other species of the families *Enterobacteriaceae* and *Vibrionaceae*. *V. metschnikovii is* generally susceptible to antibiotics.

INDEX

W

X

Y